Data Science in Practice

Data Science in Practice is the ideal introduction to Data Science. With or without math skills, here, you get the all-round view that you need for your projects. This book describes how to properly question data in order to unearth the treasure that data can be. You will get to know the relevant analysis methods and will be introduced to the programming language R, which is ideally suited for data analysis. Associated tools like notebooks that make Data Science programming easily accessible are included in this introduction. Because technology alone is not enough, this book also deals with problems in project implementation, illuminates various fields of application, and does not forget to address ethical aspects. *Data Science in Practice* includes many examples, notes on errors, decision-making aids, and other practical tips. This book is ideal as a complementary text for university students and is a useful learning tool for those moving into more data-related roles.

Key Features:

- Success factors and tools for all project phases
- Includes application examples for various subject areas
- Introduces many aspects of Data Science, from requirements analysis to data acquisition and visualization

Tom Alby has been working in the digital world since 1994, including nearly 20 years for search engines such as Lycos, Ask.com, and Google. His focus is on data-driven applications for everyday business and the development of data literacy. He is the author of several books, lecturer for Data Science and Digital Analytics at various universities, and certified project manager (PMP) of the Project Management Institute since 2004.

CHAPMAN & HALL/CRC DATA SCIENCE SERIES

Reflecting the interdisciplinary nature of the field, this book series brings together researchers, practitioners, and instructors from statistics, computer science, machine learning, and analytics. The series will publish cutting-edge research, industry applications, and textbooks in data science.

The inclusion of concrete examples, applications, and methods is highly encouraged. The scope of the series includes titles in the areas of machine learning, pattern recognition, predictive analytics, business analytics, Big Data, visualization, programming, software, learning analytics, data wrangling, interactive graphics, and reproducible research.

Published Titles

Urban Informatics
Using Big Data to Understand and Serve Communities
Daniel T. O'Brien

Introduction to Environmental Data Science
Jerry Douglas Davis

Hands-On Data Science for Librarians
Sarah Lin and Dorris Scott

Geographic Data Science with R
Visualizing and Analyzing Environmental Change
Michael C. Wimberly

Practitioner's Guide to Data Science
Hui Lin and Ming Li

Data Science and Analytics Strategy
An Emergent Design Approach
Kailash Awati and Alexander Scriven

Telling Stories with Data
With Applications in R
Rohan Alexander

Data Science for Sensory and Consumer Scientists
Thierry Worch, Julien Delarue, Vanessa Rios De Souza and John Ennis

Big Data Analytics
A Guide to Data Science Practitioners Making the Transition to Big Data
Ulrich Matter

Data Science in Practice
Tom Alby

For more information about this series, please visit: https://www.routledge.com/Chapman--HallCRC-Data-Science-Series/book-series/CHDSS

Data Science in Practice

Tom Alby

CRC Press
Taylor & Francis Group
Boca Raton London New York

CRC Press is an imprint of the
Taylor & Francis Group, an **informa** business

A CHAPMAN & HALL BOOK

First English edition published 2024
by CRC Press
2385 NW Executive Center Dr, Suite 320, Boca Raton, FL, 33431

and by CRC Press
4 Park Square, Milton Park, Abingdon, Oxon, OX14 4RN

CRC Press is an imprint of Taylor & Francis Group, LLC

ISBN: 978-1-032-50524-4 (hbk)
ISBN: 978-1-032-50526-8 (pbk)
ISBN: 978-1-003-42636-3 (ebk)

DOI: 10.1201/9781003426363

Typeset in Palatino
by Deanta Global Publishing Services, Chennai, India

Contents

Foreword

There is a plethora of data science books, courses, and online tutorials, but their focus is on the application of an algorithm in a particular programming language or a general introduction shows the first step but leaves you alone with the rest. This book is about really understanding the major algorithms and being able to apply them properly.

However, a successful data science project depends on more than an understanding of a programming language or the selection of an appropriate algorithm. "Tell me how your project starts, and I'll tell you how it will end" is a saying in the project management world, and it also applies to data science. If the problem to be solved is not clearly defined from the beginning, even the best algorithms will not change the fact that the result will not be usable. Understanding the actual need as well as communicating effectively with stakeholders require skills that go beyond programming. Data science projects are mostly business projects, and most of the work needs to be invested outside of analysis and modelling.

This is the broad scope of Data Science that this book aims to cover. Hopefully, with this book, you will have everything you need for a successful Data Science project, from the initial idea to the development of an application that will go into productive use for end users, no matter whether you are a decision maker, a Data Scientist, or just interested in the topic. This book is the result of a large number of data science projects and several semesters teaching "Data Science" at the Department of Information at the Hamburg University of Applied Sciences in Germany. I have tried to combine the best of both worlds, theory and practice.

This English version has been updated to include the latest developments such as Quarto and ChatGPT. The rise of ChatGPT & Co raises voices as to whether Data Science is still necessary at all when Microsoft integrates AI into its Office product. Of course, Data Science will not disappear, because ChatGPT & Co can do many things really well, but special cases still have to be solved with the methods presented in this book.

Also, some examples have been changed as the initial version relied on a German context. All the materials for this book can be found on my website in the section set up especially for this book, https://alby.link/datascience. You will find notebooks (a wonderful combination of text, data, and code, see chapter 5), as well as more detailed information.

I would love to get your feedback, please don't hesitate to send me an e-mail to tom@alby.de. I am particularly interested in what helped you, but also what didn't, where you had difficulties in understanding, for example. I don't always answer quickly, but eventually I do.

I have to thank my family, Anke, David, Emilia and Glenn, who had to do without me for many hours while I was working on this manuscript. Writing a book like this is not a relationship-building act. Thanks also go to the students who gave me feedback on my lecture notes or simply served as "guinea pigs" when I tried out whether one explanation of complex issues worked better than another. I teach mostly students outside of STEM subjects, and I know that it takes a lot of courage to come to my course. I would also like to thank my test readers Dr. Meltem Akyilmaz-Yabaci, Vincent Domanski, Dr. Christian Messerschmidt and Dr. Gerrit Vorjans. They worked through early versions of the manuscript and discovered many errors and inconsistencies. All errors that are still present in this book are mine alone.

Figures

1

Introduction

We are drowning in data but starving for insights. (Author unknown)

1.1 The Age of Data: Is It Just a Hype?

Data is the new oil – at least, that's what many PowerPoint slides say. Whether the analogy between data and oil is correct is debatable. Unlike oil, data is not consumed when it is used. Data can even be improved by its use and the new data it generates. A famous example of this is the Google search engine: the fact that billions of people use the search engine every day means that Google's systems learn which search hits are apparently perceived as more relevant, using the resulting data to further train the ranking algorithms. It's probably legitimate not to take the comparison between oil and data too literally, but to interpret it in terms of what oil did for industrialization and what data will do in the digital age.

At the same time, there is no denying that a new technology or even a technology term in the past was greatly overrated at the beginning, and thereafter there was a deep fall into the "valley of disappointment," as described in the Gartner hype cycle [27], see Figure 1.1. Comparing the share prices of technology stocks between 1995 and 2010 to the hype cycle, we see the peak of exaggerated expectations with the dotcom boom. The Internet, which had just become popular thanks to the World Wide Web, fueled the imagination of investors, so that money was also invested in ideas for which the infrastructure at the time was not yet mature [3]. This was followed by the dotcom crash, with shares plummeting. In the meantime, however, the values have recovered, because the Internet was not shut down after the crash. Today, it is impossible to imagine life without it.

The artificial intelligence world has already gone through two peak phases, as you will learn in Chapter 2, and both phases were followed by an *AI Winter* in which research funds were withdrawn. In both cases, no "path of enlightenment" was found, nor was a "plateau of productivity" reached. Why should it be different this time?

DOI: 10.1201/9781003426363-1

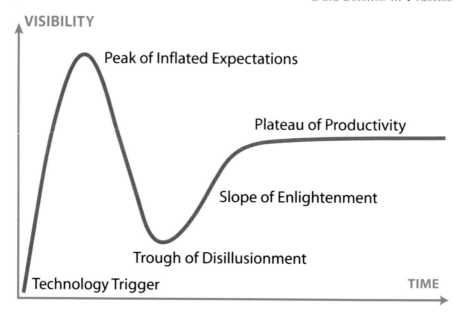

FIGURE 1.1
The Gartner hype cycle, source: Jeremy Kemp, CC BY-SA 3.0.

1.2 Why Is Data Science Relevant Now?

The possibilities for capturing and processing data have changed rapidly since the early days of electronic data processing. This is due not only to the ever-growing computing and storage capacity, but also because of the increasing networking of computers, cell phones, sensors and many other devices that have access to the Internet. Today, we can even control vacuum cleaner robots with an app via the Internet, and kitchen machines transmit data about our cooking behavior to the manufacturer. The convergence of these developments is leading to a new starting point for artificial intelligence applications. Having said that, the term *AI* is often used when it is not appropriate.

However, the term "Data Science" fits very well, because the large amount of data creates new challenges: No human being can handle these huge amounts of data and recognize patterns in them as a machine can. If data is the new oil, then data science is the science of how to produce something from the raw material that generates added value. Some even talk about artificial intelligence being the new electricity.

Data and the capability of exploiting it have paved the way for new business models:

- Airbnb has more rooms on offer than the world's largest hotel chain, InterContinental Hotels Group [37]. At the same time, Airbnb doesn't own a single one of those rooms. Airbnb's greatest treasure is its data, user-generated reviews, user-created descriptions, and a platform that searches available rooms around the world. Anyone can become a landlord from one moment to the next, barriers to entry have fallen, and even prices can be set dynamically using machine learning models, depending on supply and demand. There is no need for a reception desk or hotel management.

- The company Kobalt is not a household name to most people, but for professional musicians, this service is a popular platform to get transparency about the usage of their works [29]. Instead of waiting months or even years for the record company to produce a barely comprehensible statement and pay the artist, via Kobalt, musicians can see in real time how many of their works are currently being sold on CD, used in a commercial, or streamed on Spotify. Illegal uses are also automatically identified and money claimed for them. And instead of having to go through a complex bureaucratic process, interested parties can use the platform to request rights for a piece of music, for example to use it in an event, and the artist can release the rights at the click of a mouse. A record label is hardly needed here, with Kobalt even offering a label itself. Various machine learning technologies are used here, not only to be able to process the large volumes of data (Kobalt reportedly creates more than 1,000 data points for each song) but also to identify unauthorized uses.

- Google achieves much of its revenue through Google Ads, formerly AdWords, a platform through which advertising is sold. Instead of hiring salespeople to sell the ad space, a self-optimizing system does the work, because Google has its own sales department only for large customers. Advertisers sign up on the platform, create ads and specify which keywords they want to appear for. Then, they set a bid for a click on that ad. Supply and demand do the rest. The more advertisers want to be found for a search term, the higher the prices go in the auction, and the more Google earns. And if an ad is apparently not that relevant to a search query and is therefore hardly clicked on, the advertiser automatically receives a "malus" for the combination of the search term and the ad, i.e. the advertiser has to pay more. Thousands of these auctions take place every second. Without people involved in the sale. Without manual pricing.

However, these success stories should not hide the fact that there are also failures, which will also be mentioned in this book. With regard to Data Science, in the Hype Cycle for Analytics and Business Intelligence 2019, Data Lakes are are already on the verge of the valley of disappointment, whereas Data Literacy is just on its way to the peak of inflated expectations. This may be due in part to "overselling" (as in the IBM example in Chapter 2), or perhaps because the prerequisites for a successful Data Science project are not in place. More on this in Chapter 3, "Process of a Data Science project".

1.3 Why Data Science with R?

My passion for R started in the early 2000s when I was looking for a free software for data analysis and visualization. It was love at first sight, even though I've written plenty of Python code since then, in addition to countless R programs. R is more in line with the way I think. I am sure that you will fall in love with R as well.

In teaching practice, I have realized that the introduction to R is a lot faster than the introduction to Python, enabling me to get to the Data Science topics much faster. Data analysis is already hard enough for some, and then to get an introduction to programming ... In connection with R, most students do not even realize that they are just learning programming in addition to statistics.

I usually need half a day for the introduction to R (see Chapter 4), the software library Tidyverse, which is popular in the data analysis field, and notebooks (Chapter 5). Then, in the afternoon, students are introduced to data analysis (also Chapter 5). RStudio is also a wonderful integrated development environment that greatly simplifies data analysis. Such a tool, especially designed for data analysis, can be found in the Python world so far only in rudiments (for example in PyCharm). R is also much easier to install than Python, although kaggle.com and other platforms for both languages offer a quick start via notebooks.

However, this does not mean that Python is less suitable for data science. It is not uncommon for R to be used to first analyze the data and build initial models. This is exactly what we do in Chapters 6 to 9. For use in a productive environment, Python is often used, although this would also be possible with R. So it's not an either/or but a both/and. And those who have learned R can learn Python all the more quickly.

1.4 Who Is This Book For?

I haven't written this book for prospective computer scientists, but primarily for those who want to approach Data Science without having previously completed an advanced math course, even if they are not planning a career as a Data Scientist. Data Science will most likely be with us for some years to come, and therefore, it is immensely important that Data Scientists and stakeholders understand not only the possibilities but also the implications of Data Science (see also [6]).

For those who want to go deeper into Data Science, this book offers a compact introduction with examples and tips for further exercises and next steps on the way to becoming a professional Data Scientist. Chapters 10 to 12 offer you additional content beyond the introduction to R and pure algorithms.

For decision makers who want to have a holistic view on Data Science, this book offers many "best practices" that you should consider for starting your own Data Science projects. Reading it will also help you find suitable candidates (for example, use the glossary to find out if someone knows what *heteroscedasticity* means).

1.5 Is It Possible to Learn Data Science without Math?

This is where opinions differ. Some experts think that Data Science cannot work without a degree in mathematics or at least something related to mathematics. Others say that knowledge of mathematics is only necessary for very complex issues and that one can get very far with the standard methods.

In my opinion, both sides are right, but even with the standard methods, you can get a lot wrong if you don't understand the basic principles of the respective algorithm, and for that you need some mathematics.

This book is also not entirely without math. But, the experience of many years of teaching has shown that math can also be presented in such a way that it does not hurt. Quite a few participants come to my courses with the conviction that they can't do math and never will. The mere thought of a formula frustrates them and causes them to switch off completely. Sometimes, bad experiences from school days are responsible for this unnecessary phobia. The reason students still come to my course is because I promise that there will be as little math as possible. And then, I introduce the necessary principles in such a way that they hardly realize they've just done anything with math. It's a little like going to the dentist: if I see the torture instruments as soon as I walk in, I get a bad feeling, at least for me. If I only see the "usual" cutlery, then my pulse doesn't go up at all. So, if mathematics also basically evokes rather negative feelings in you, then I ask for your trust that I can help you to understand at least the basics that you need for an elementary understanding of the common Data Science approaches.

According to the Harvard Business Review, the job of a Data Scientist is "the sexiest job of the 21st century" [20]. In the age of data, it is the Data Scientists who can handle the huge masses of data, find patterns in it and advise senior management on what to do next. However, it is difficult to find this "one-man band" because, as we will see in the definition of the individual areas of artificial intelligence, a successful data science project does not require an excellent grade in math alone. And that's exactly why there are opportunities in the data science field for those who don't have a PhD in mathematics. You may not become the top Data Scientist without a math degree. But at least, after reading this book, you won't blindly throw any data at an algorithm just because it wasn't explained better in a YouTube tutorial. And you may have a special talent for storytelling, which many Data Scientists lack.

1.6   How to Use This Book

Should you simply want to understand what Data Science is exactly and how and where Data Science can be applied, it is not necessary that you recreate every programming task in this book yourself or read and understand the program code completely. The book is structured in such a way that you can skip every line of code, but still have plenty to read. However, I still encourage you to try out the code examples, because I promise that just understanding them will not only make you happy, but also give you a deeper understanding of how Data Science works.

You can read the book from cover to cover or just read the chapters that interest you. If I presuppose knowledge from another chapter, I mention this with a reference to the respective section.

2

Machine Learning, Data Science, and Artificial Intelligence

[...] I believe that at the end of the century the use of words and general educated opinion will have altered so much that one will be able to speak of machines thinking without expecting to be contradicted. [49]

Although the term "Data Science" was coined in the 1960s, it initially had nothing to do with artificial intelligence, which had first gained popularity a few years earlier. The foundations for machine learning emerged much earlier and laid the foundation for many artificial intelligence concepts, but were initially based simply on statistical models. How exactly the domains differ is the topic of this chapter.

2.1 Learning from History – All Just Hype?

Anyone who uses the term "artificial intelligence" (AI) risks being accused of using a buzz word. No wonder, because the history of artificial intelligence is still marked by many disappointments. In this chapter, we examine what is different this time.

2.1.1 Data and Machines before the Dawn of AI

Data is not a recent phenomenon. Even thousands of years ago, data was collected, if only to count something and store the result. Farmers who wanted to count their livestock used stones, one for each animal, and put them in a bag to be able to recount at a later time. People were also counted. The most famous of all censuses in the Christian West is already mentioned in the Bible, and all of them had the same goal, namely, to estimate the people in order to be able to calculate potential tax revenues. At that time, the abacus was already widely used as a calculating machine.

In the 17th century, work on mechanical calculating machines began in various places by several people. Schickard, Pascal, and Leibnitz (see [52]). In 1646, the term "data" was used for the first time in English for given quantities in mathematical problems [43].

DOI: 10.1201/9781003426363-2

What Exactly Is Data?

Data is derived from the Latin word *dare* (to give). The connotation changed in the course of the 18th century from the given, i.e. a fact, to the result of an experiment or a calculation. The word "date" also derives from *dare* [43].

The dream of a machine that can think for itself inspired people early on and also ensured the success of the Mechanical Turk, a machine built in 1769 that appeared to function like a chess robot. In fact, however, the machine contained a small human being. Amazon was inspired by the Chess Turk to name its crowdsourcing service "Amazon Mechanical Turk."

The Arithmomètre by Charles Xavier Thomas (1820) was the first mass-produced and widely used calculating machine. It was based on Leibnitz's machine. A few years earlier, the first machines capable of performing tasks automatically had been created. The Jacquard loom worked on the principle of a punched card on which a pattern was "pre-programmed". One could also understand the pattern on the punched cards as a data set, but at least software and hardware were separated from each other, so to speak. The Jacquard loom made it possible to produce faster, but also with fewer personnel, which led to the weavers who had previously worked in the traditional way fearing for their existence. Obviously, this is not the first time that jobs have been threatened or even eliminated by machines. For this is precisely what is also feared from the introduction of AI systems.

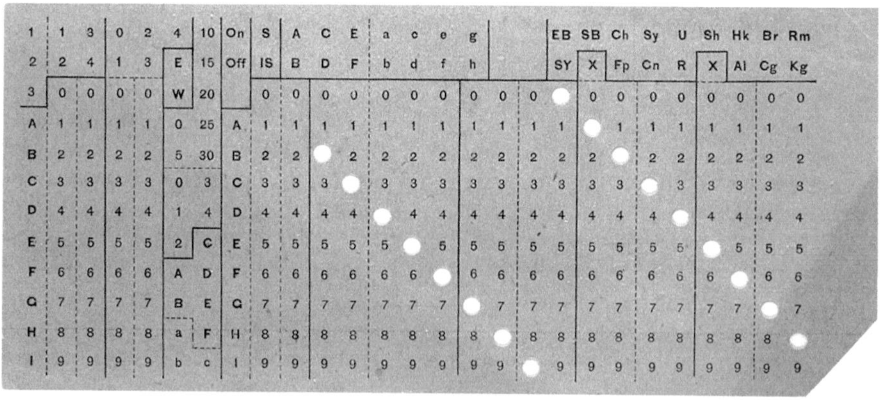

1	1	3	0	2	4	10	On	S	A	C	E	a	c	e	g		EB	SB	Ch	Sy	U	Sh	Hk	Br	Rm
2	2	4	1	3	E	15	Off	IS	B	D	F	b	d	f	h		SY	X	Fp	Cn	R	X	Al	Cg	Kg
3	0	0	0	0	W	20		0	0	0	0	0	0	0	0	0	0	0	0	0	0	0	0	0	0
A	1	1	1	1	0	25	A	1	1	1	1	1	1	1	1	1	1	1	1	1	1	1	1	1	1
B	2	2	2	2	5	30	B	2	2		2	2	2	2	2	2	2	2		2	2	2	2	2	2
C	3	3	3	3	0	3	C	3	3	3		3	3	3	3	3	3	3	3		3	3	3	3	3
D	4	4	4	4	1	4	D	4	4	4	4		4	4	4	4	4	4	4	4		4	4	4	4
E	5	5	5	5	2	C	E	5	5	5	5	5		5	5	5	5	5	5	5	5		5	5	5
F	6	6	6	6	A	D	F	6	6	6	6	6	6		6	6	6	6	6	6	6	6		6	6
G	7	7	7	7	B	E	G	7	7	7	7	7	7	7		7	7	7	7	7	7	7	7		7
H	8	8	8	8	a	F	H	8	8	8	8	8	8	8	8		8	8	8	8	8	8	8	8	
I	9	9	9	9	b	c	I	9	9	9	9	9	9	9	9	9		9	9	9	9	9	9	9	9

FIGURE 2.1

A Hollerith punch card from the 19th century.

The U.S. census around 1890 marked a turning point in the history of data processing, as the principle of punched cards was used for the first time to speed up data counting. You can see an example of a punched card in Figure 2.2.

The Origin of IBM

The German-born Hermann Hollerith had developed his tabulating machine in such a way that it was possible to analyze the data within weeks, instead of the several years it had previously taken. Hollerith's company later became IBM, an acronym for International Business Machines. In addition to tabulating machines that could evaluate punched cards, office equipment such as hole punchers were available.

Thomas Watson Senior, the CEO, allegedly said that there was only a need for five computers in the entire world. In fact, however, it could not be proven that he actually said that. However, his son Thomas Watson Jr. seemed to be more convinced of the potential of computers and thus changed IBM permanently. His book *The Father, the Son and the Company* describes how high a risk was that the junior Watson took with the company at that time (see especially [53]).

Tabulating machines were used until the early 1960s, by which time punched tapes were faster and in turn, were overtaken by magnetic data recording.

In the following decades after Hollerith, development accelerated. Konrad Zuse built the Z1, the first "mechanical brain" in the second half of the 1930s; the Z3 that followed in 1943 is considered the world's first functional electronic computer based on relays. Based on the work of Turing during the Second World War, which led to the decoding of German Wehrmacht messages, the ENIAC, the first digital programmable tube computer, was developed in 1946 (see Figure 2.1).

In the 1950s, the first commercial computers appeared on the market, including the UNIVAC I and the TRADIC. The latter was the first computer to use transistors instead of vacuum tubes. Also, in this decade, the IBM 1401 was introduced, which is considered the forefather of modern computing. The monthly rent for such a computer, adjusted for inflation, was 23,500 dollars, a hefty price, but one that nevertheless brought productivity gains to many companies. DEC introduced the PDP-8 in the 1960s for less than 20,000 dollars. Adjusted for inflation, that's still just under 1,000,000 dollars today, but still, the machine was so affordable that more and more research institutions could afford computers.

FIGURE 2.2
The ENIAC, Source: US Army (Public Domain).

2.1.2 The First Spring of Artificial Intelligence

This development of hardware fueled the imagination of scientists and the public. In the 1950s, researchers were confident that it would only take a decade before a computer could beat a human at chess. This phase, known in retrospect as the "spring", was full of euphoria. In fact, however, it was not until 1997 that IBM's Deep Blue was able to beat the then world chess champion Garri Kasparov.

Many research projects were driven by the American military, for example, they wanted to be able to translate Russian documents automatically. Today, we already see good programs like Deepl.com or Google Translate, but at that time they were still a long way off. The concepts for neural networks were also developed at that time.

Another approach was the General Problem Solver by Simon and Newell, which was developed as early as 1957. This software was supposed to divide a problem into smaller subproblems, which were then easier to solve.

2.1.3 The First AI Winter

In the early 1970s, the first AI winter broke. An AI winter is the term used to describe phases in which research funding for AI projects is withdrawn or severely curtailed because the hoped-for results fail to materialize. This first AI winter was triggered by several publications, including a book on perceptrons by Minsky and Papert, which showed the limitations of the approach.

The General Problem Solver had been discontinued earlier. The Defense Advanced Research Projects Agency (DARPA) subsequently cut funding, making it difficult if not impossible to obtain research funding for artificial intelligence projects in the 1970s. A *bon mot* from this period is the translation of one of the systems of the time from English to Russian and back again. "The spirit is willing, but the flesh is weak" became "The vodka is good, but the flesh is rotten." After many years of high spending, the disappointment of the backers of the time is understandable.

2.1.4 The Second AI Spring: Expert Systems

After the failure of the general-problem-solver approach, the focus at the end of the 1970s was on systems that were to be fed with so much knowledge, at least in one field of knowledge, that they were capable of making conclusions independently. These so-called expert systems became popular in the 1980s and were again able to attract research funding.

One of the languages used for this purpose was Lisp which ran on purpose-built Lisp machines. These computers were replaced after a few years by cheaper general-purpose computers such as those from Sun, so initially this industry collapsed, but the hype around expert systems continued. Besides Lisp, Prolog was popular. The name is composed of PROgramming in LOGic (see Figure 2.3).

Both languages still exist today, although they no longer experience the popularity that R, Python, and Julia do today.

```
Welcome to SWI-Prolog (threaded, 64 bits, version 9.1.11)
SWI-Prolog comes with ABSOLUTELY NO WARRANTY. This is free software.
Please run ?- license. for legal details.

For online help and background, visit https://www.swi-prolog.org
For built-in help, use ?- help(Topic). or ?- apropos(Word).

?- [user].
|: hello :- format('Hello world~n').
|:
% user://1 compiled 0.01 sec, 1 clauses
true.

?- hello.
Hello world
true.

?- |
```

FIGURE 2.3
An example of Prolog programming.

2.1.5 The Second AI Winter

The expert systems did not get beyond a spring either, because they did not meet expectations. One major criticism was that the systems did not learn independently and were therefore expensive to maintain. Another criticism was that some grotesque responses occurred when the expert systems were not questioned exactly as intended. Thus, toward the end of the 1980s, the second AI winter began, this time lasting longer than the first.

2.1.6 Is This a New AI Spring?

In 2011, IBM Watson, the successor to IBM's Deep Blue, won Jeopardy!, a game that until then had been thought to be possible to won only by humans. That's because Jeopardy! doesn't seek answers to questions but gives answers for which the questions are sought. This sensation further boosted hopes that the artificial intelligence code had finally been cracked. In 2016, Google introduced AlphaGo, an artificial intelligence that could beat humans at the game of Go. Go is a lot more complex than chess, and previous programs were at most beginner-level. AlphaGo was followed by AlphaGo Zero, which was given only the rules of the game and then trained against itself. After AlphaGo Zero came Alpha Zero, another AI that was not designed specifically for the game of Go, but could learn various games based on the rules and games against itself and beat all other previous systems as well as human players.

Google's DeepMind products are not an indication not only that the third AI spring has not only begun, but also that there will no longer be such a harsh winter. For example, while during the first AI spring phases, software was mainly written in labs and otherwise kept secret, today many free software libraries exist. In 2015, Google released TensorFlow, a machine learning framework, as open source software, making it open to virtually anyone interested in learning about the libraries used by Google for speech recognition and search.

At the same time, we now have access to much more powerful computers than ever before. Every iPhone has more computing capacity than the computer that flew the first men to the moon. You can buy your own computer cluster for just a few hundred euros. The democratization of software, as well as the availability of computer power virtually for everyone, will ensure that the topics of machine learning and artificial intelligence will not disappear so easily again.

2.1.7 Setbacks and New Hopes

Nevertheless, there are setbacks in this third spring as well. After Watson's victory on Jeopardy!, IBM TV commercials were aired for Watson, showing how all sorts of problems can be solved by AI. In reality, however, it wasn't that easy. Among other things, Watson was shut down at a cancer treatment clinic after the AI made "unsafe and incorrect treatment recommendations" [14].

On the contrary, *Large Language Models* (LLM), such as ChatGPT, which gained significant popularity in November 2022, have led to a new hope in the field of AI. The run on ChatGPT is a good indicator how AI has become "tangible" to everyone, in particular due to its integration into the Bing search engine, allowing all interested parties to "converse" with artificial intelligence.

LLMs comprise billions of parameters in a neural network. Such models are trained with vast amounts of text based on self-supervised or semi-supervised learning. In contrast to the previous distinction between supervised and unsupervised learning, self-supervised learning doesn't require labeling. The computer identifies "signals" for supervised learning from the large volumes of text. Another difference from previous approaches is that LLMs can be used for general purposes rather than having been trained for a specific task; they are essentially pre-trained for various purposes, hence the "P" in "GPT". For instance, it is possible to perform a sentiment analysis without the model having been previously trained to understand what could be meant positively or negatively. Conversations with a LLM can be very convincing, despite their "hallucinations", i.e. facts being made up. Nevertheless, Google's LaMDA was even thought to have a mind [38].

2.1.8 Technological Singularity: Do Machines Have a Mind?

The concept of the singularity became popular through Ray Kurzweil's books, especially [34]. His hypothesis is that there will come a point when machines will be just as intelligent as we are and will then overtake us. This will allow the machines to improve themselves, which at the same time would make the future of humanity uncertain. The fear that an artificial intelligence could dominate mankind is part of many novels and movies, among others *Ex Machina* by Alex Garland or *Blade Runner* by Ridley Scott and its second part by Denis Villeneuve.

About Ray Kurzweil

Ray Kurzweil is a computer pioneer whose inventions are used everywhere today, but they are not always associated with his name. Kurzweil is often credited as the inventor of Optical Character Recognition (OCR), although in fact, systems had existed earlier that could recognize text. Kurzweil's credit here is that he built a system that could read text aloud to blind people. One of the first customers was Stevie Wonder, which led to a personal contact. Together, they developed the Kurzweil synthesizer, which in the 1980s could mimic instruments such as a piano so faithfully that musicians had trouble distinguishing the synthesizer from the real instrument.

Kurzweil has been working at Google since 2012, where his task is the understanding of natural language by machines.

Fiction undoubtedly also has an influence on reality. A few years ago, for example, a rumor arose in social networks that Facebook had to shut down two artificial intelligence bots because they had invented their own language. This never happened in that way, but it also shows people's fears [36]. Nevertheless, it is assumed that the time of the technological singularity will come as a surprise, even if it has already been postponed into the future several times (see above all [55]).

Another exciting question is whether an artificial intelligence could then also have a mind. An interesting thought experiment on this was proposed by the philosopher John Searle in 1980. In a room, there is a person who receives instructions in Chinese, which he does not understand but can follow with the help of rules and thus formulate a response in Chinese. The people outside the "Chinese Room" think that the person in the room speaks Chinese, although this is not the case. Searle wanted to prove that a machine need not actually be intelligent just because it passes the Turing test (see next section), because it can do so without understanding the meaning of what it is doing.

Searle's thought experiment has not been received without criticism. One question is whether it is even necessary for an artificial intelligence to have a mind.

2.1.9   Alan Turing and the Turing Test

Alan Turing not only developed the theoretical basis of today's computer, but also provided a test as early as 1950 that is still considered useful today for checking the existence of artificial intelligence. This test works like this: a test person has two interlocutors with whom he communicates via a keyboard. One of the two interlocutors is a human, the other a machine. If the test person cannot determine which of the two interlocutors is the machine, the Turing test is passed. Turing himself believed that by the year 2000, this test would be such that test subjects would only be able to distinguish a human from a machine with a probability of 70%.

About Alan Turing

The fact that Alan Turing's work helped to decipher the secret codes of the German Wehrmacht and that his thoughts are the basis of today's theoretical computer science did not protect him from being persecuted because of his homosexuality. Thus, in 1952, he was sentenced to therapy, as homosexuality was still a crime at the time. The therapy caused Turing to suffer from depression, which led to his suicide in 1954. In 2009, then-Prime Minister Gordon Brown apologized for the treatment of Turing, but it took until 2013 for Queen Elizabeth to posthumously pardon him.

Until now, the Turing test has not been passed. However, there are weakened variants of the Turing test that certainly produced winners.

2.2 Definitions

Machine learning, artificial intelligence, statistics – what does all this have to do with Data Science? In this section, we will review different interpretations of the terms.

2.2.1 Machine Learning

[31] describe that machine learning

> focuses on the design and evaluation of algorithms for extracting patterns from data. Data mining generally deals with the analysis of structured data and often implies an emphasis on commercial applications.

The difference from data mining is thus determined here by the fact that data mining deals with structured data, whereas machine learning does not. [7] defines machine learning as follows:

> Machine learning is about extracting knowledge from data. It is a research field at the intersection of statistics, artificial intelligence, and computer science and is also known as predictive analytics or statistical learning.

This definition includes the overlap with statistics, although the vocabulary of statistics differs in part from that of machine learning. For example, in statistics, we speak of independent variables or predictors whereas in machine learning, we speak of features. Despite the language differences, the closeness to statistics cannot be denied and is also a frequent topic of memes that aim to show that machine learning is just old wine in new skins. However, in machine learning, a greater influence of computer science can be observed. At the same time, the definition does not see Machine Learning as part of artificial intelligence but merely attests to an intersection.

2.2.2 Artificial Intelligence

The field of artificial intelligence is somewhat older, as discussed in the previous section. Nevertheless, there are different phases. On the one hand, there is the hope or even fear that a strong artificial intelligence will catch up with us or even overtake us. HAL 9000 from the movie *A Space Odyssey* by Stanley Kubrick is an example of such a strong AI. On the other hand, today we see the results of so-called weak artificial intelligence, such as Alexa, Apple Siri or the Google Assistant: systems that have been trained for a specific purpose and can do it more or less well. We can thus say that machine learning is part of the field of artificial intelligence.

2.2.3 Data Science

First of all: There is no universally accepted definition of Data Science. This may seem strange, but this field is still very young, so not all details have been defined yet. In fact, there are very different views on what Data Science is and what it is not.

Almost all of the following definitions imply that Data Science can be used to identify and exploit complex patterns in data. Therefore, right up front: Of course, simple patterns can also be extracted from data, but if a human can identify such patterns, then it is no longer necessary to employ a machine to do so. We are talking about cases where the patterns in the data are not necessarily obvious. This can mean either that it would take a human too much time to work through a certain amount of data. Humans are good with a few attributes (attributes are used synonymously with features or variables in this book) but as soon as there are more of them, it becomes difficult for a human to see connections.

In Davenport and Patil's famous *Harvard Business Review* article "Data Scientist: The Sexiest Job of the 21st Century," Data Science itself is not defined, but the job of a Data Scientist is described:

> Data scientists are the key to realizing the opportunities presented by big data. They bring structure to it, find compelling patterns in it, and advise executives on the implications for products, processes, and decisions. They find the story buried in the data and communicate it. [20]

[42] define Data Science in their terrific book "Data Science for Business" (2013) as follows:

Data Science involves principles, processes, and techniques for under-standing phe- nomena via the (automated) analysis of data.

[31], on the other hand, differentiate that while Data Science makes use of other fields such as machine learning, overall it is "broader in scope" as it addresses other challenges. These include the collection, cleaning, and trans-formation of unstructured data from the Internet, the use of Big Data tech-nologies, and the ethical reflection of data use.

Another definition results from Figure 2.4. Business value plays a more important tole here, but there is also a field for more legal issues that no Data Scientist can duck away from (see especially Chapter 11, "Ethical Handling of Data and Algorithms"). The size of the circles in the figure also indicates that coding is only part of the actual work.

My definition, thus, is as follows:

Data Science is currently a still consolidating application area of arti-ficial intelligence, in which interdisciplinary knowledge from statistics, computer science, and expert knowledge from other disciplines are con-verged to further develop methods for unlocking data patterns in these fields.

Data Science makes use of various sciences, but is not (yet) a science in itself. This may change in the future.

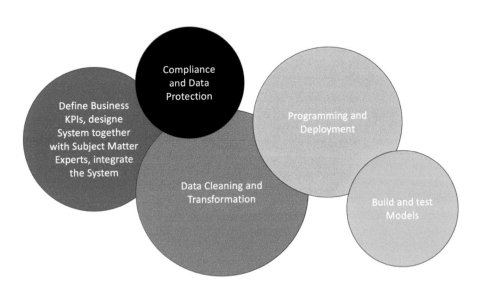

FIGURE 2.4
Proportions of the different tasks within a Data Science project.

2.2.4 Data Analysis and Statistics

Data analysis and statistics are not the same; quite the contrary. [44] formulates the difference in such a way that in statistics, the focus falls on mathematical derivations, but data analysis has a stronger application focus. This does not mean that there is no use of statistical methods in data analysis: quite the opposite. Ask a statistician whether a rank correlation or a product-moment correlation coefficient is used, and he will explain the mathematical background. In data analysis, you are more likely to have the use cases explained to you.

Data analysis (with knowledge of statistical methods) is a subtask in machine learning and therefore also in Data Science projects. You can't march right through to the machine learning algorithms, because first you have to understand your data based on an analysis.

2.2.5 Big Data

Even though there is a lot of talk and writing about Big Data, there is not much of it to be seen in the wild, if it goes by the definition that three criteria must be fulfilled for this (the three Vs):

- Volume – For example, how much data must be processed

- Variety – For example, whether different types of data, structured and unstructured, need to be merged

- Velocity – For example, the speed of change

In the Gartner hype cycle, the concept of Big Data is already past its peak.

Google's search engine is a good example of Big Data: huge amounts of data are collected by crawlers, HTML files, images, videos, links within the documents necessary to build a web graph, and spam defense, and since the Web is constantly changing, high processing speed is necessary, so that the most important changes are also quickly visible in the index.

As a counterexample to Big Data, the German version of Wikipedia can be downloaded as an XML dump; about 20 gigabytes of hard disk space is occupied by this file. I can process this file on my small MacBook Air. It contains more or less well-structured data with references to multimedia files, and if I were to get the Wikipedia changes in real time, still, not all three Vs would be fulfilled. Speed could still be an issue, depending on what I plan to do with the data, but within a couple of hours, I can create a classifier based on Wikipedia data using R.

Nevertheless, today's data volumes, even if they do not meet all three Vs, have led to the emergence of some frameworks that help manage them. For example, it is no longer necessary to purchase a dedicated cluster; if needed, it can be rented from a cloud provider just for the time it is needed. Here, programming models such as MapReduce are used for processing distributed data. In Chapter 10, "Workflows and Tools," we look at Apache Spark, an extension of MapReduce.

There will also be more applications that need to process Big Data. If we think of all the sensors that are now Internet-enabled, for example in self-driving cars, then vast amounts of data are already being collected that can lead to new applications.

3

The Anatomy of a Data Science Project

Tell me how your project starts, and I'll tell you how it ends. (Wisdom from the project management world, author unknown)

In this chapter, we will cover everything that is necessary to manage a Data Science project. Almost every Data Science project attempts to solve a problem that exists in the real world, and in order for this ambition to succeed, a number of project phases and best practices have been established. This starts with requirements analysis, and touches data acquisition and cleaning, the soft factors of a model's impact, and finished with testing and rolling out a project.

3.1 The General Flow of a Data Science Project

Even though Data Science is still a young field, some best practices have already been established. Some of them originate from related fields such as data mining. We first look at processes that are mentioned in the (partly still young) literature and then take a closer look at fundamental patterns in the further sections.

3.1.1 The CRISP-DM Stages

The acronym CRISP-DM stands for CRoss-Industry Standard Process for Data Mining, which was developed by a consortium within an EU-funded project starting in 1996 (see Shearer 2000 [45]). This process includes a granular overview of the individual steps, which I supplement here with references to where you can find these steps in this book:

1. Understanding the business and its operations (this chapter)

   - Determine the business objectives for this project
   - Evaluation of the current situation
   - Determine the goals for the data mining (or the Data Science project).

DOI: 10.1201/9781003426363-3

- Creating a project plan

2. Understanding the data (see Chapter 5 on exploratory data analysis.)

   - Collect first data
   - Describe data
   - Explore data
   - Checking the data quality

3. Preparation of the data (also Chapter 5)

   - Select data
   - Clean data
   - Transform data (in English construct, but here I interpret it as transformation)
   - Integrate data
   - Format data

4. Modeling (Chapters 6 to 9)

   - Select model technique (here, transferred to Data Science, the algorithm is meant, whereby several algorithms and models are usually tested against each other)
   - Generate test design
   - Create model
   - Rate model

5. Evaluation (this chapter and Chapters 6 to 9)

   - Evaluate results
   - Check process
   - Determine next steps

6. Deployment (use of the system in production: see this chapter as well as Chapter 10)

   - Planning the deployment
   - Planning the monitoring and maintenance of the model
   - Creation of a final report
   - Project review (for example, post mortem, lessons learned)

Even though this looks like a linear process, Figure 3.1 shows that the steps in the CRISP-DM are not processed as in a waterfall model but that iterations may be necessary. For example, after the evaluation, you may have to go back to understand the business all over again because completely new facts have emerged from the data. Is that rare? Not at all, but it clearly depends on how much time you have invested in this first phase.

FIGURE 3.1
The CRISP-DM model (according to the IBM SPSS Modeler CRISP-DM Guide, CC BY-SA 3.0).

3.1.2 ASUM-DM

ASUM-DM stands for Analytics Solution Unified Method for Data Mining and was released by IBM in 2015 as an evolution of IBM's CRISP-DM. It comprises five phases:

1. Analysis
2. Design
3. Configuration and production (Configure and Build)

4. Commissioning (Deploy)

5. Operation and Optimization (Operate and Optimize)

Here, too, the phases can be run through several times. A project management stream runs in parallel. A detailed description can be found at https://alby.link/asum.

3.1.3 The Data Science Workflow According to Hadley Wickham

Hadley Wickham, Chief Data Scientist at Posit PBC, creator of the Tidyverse, and multiple book author, has developed his own flow, which you can see in Figure 3.2. His process focuses more on the Data Scientist's analysis activities and completely excludes the business view.

His slightly different view does not mean that his approach is not suitable for business-related projects; he simply has a different perspective, a different focal length in his lens, so to speak, through which he sees a particular section of a Data Science project in more detail. This view is underpinned by his R package Tidyverse, which will be covered in Chapter 5.

Wickham's process consists of:

- Importing the data

- Cleaning the data

- Data transformation

- Data visualization

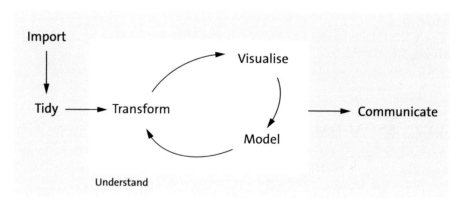

FIGURE 3.2
Hadley Wickham's flow of a Data Science project, reprinted with his kind permission.

- Modeling

- Communication

He additionally characterized the transformation, visualization, and modeling steps as a process of understanding. While the CRISP-DM approach first presupposes the understanding of the data, Wickham assumes that a more precise understanding can only arise during transformation, visualization, and modeling. I can only endorse this view, as it is only upon closer inspection that further questions arise. However, this is the view of the Data Scientist.

3.1.4 Which Approach Is Right for Me?

The answer to this question depends on the project, but also on the organization. As discussed in the previous section, Wickham's approach is most suitable for Data Scientists themselves, but even a Data Scientist cannot turn a blind eye to the needs of the organization. The CRISP-DM approach has the whole organization in mind as well as the steps outside the Data Scientist's remit. My recommendation is a blend of both approaches, though I recommend Wickham's approach primarily for exploratory data analysis and model building, so everything you'll find here is in Chapters 5 through 9.

3.2 Business Understanding: What Is the Problem to Be Solved?

Before you start developing models, you must first understand what the problem is that needs to be solved and what the user expectations are.

3.2.1 Senior Management Support and Involvement of the Specialist Department

Data Science projects usually fail for two reasons (see also Alby, 2017 [4]):

- The requirements were not properly understood, or the problem was not properly solved

- There was a lack of support from management and the technical department.

This is exactly what is needed from the stakeholders that can make a Data Science project fail, the senior management (board level, executive, and management) and the business department. The business department can lose interest for a wide variety of reasons, be it because everything is more complicated than expected and you also have to collaborate more than expected, or because there are latent feelings of disturbance because something is changing.

By lack of support, however, we mean not only during the project but also during the time after a Data Science application has been made available to users. The best systems in the world can be created, but they are of no use if they are not used. This also needs to be discussed from the beginning: How can it be ensured that the system being created will have appropriate usage? This is only possible if the hoped-for benefit is then actually present. Understanding exactly what this benefit consists of is what the next section is about.

3.2.2   Understanding Requirements

The biggest challenge in a Data Science project is often not training and optimizing a model but rather, understanding what it is actually about and how value can actually be generated (not to mention getting the right data and cleaning it). This also involves understanding the business model in which the users operate, which includes understanding the goals of the department in question.

Example: At the beginning of the Corona crisis, I was working on a Data Science project for a debt collection company. I had no prior knowledge of debt collection, and in fact, debt collection works very differently than most people probably imagine. The world of debt collection is much more complex and exciting than expected, but it took several weeks to understand it. It is necessary to understand exactly how users work and how the proposed tool will support them. This can often require being on site watching staff work, observing them, and asking questions.

But, it's not just we data scientists who have to learn. Experience shows that users often want functions that they don't use afterwards, or that they want functions that they can't use at all because they don't make sense. It may seem counterintuitive, but sometimes something completely different should be done than what users want. My favorite example of this is the first Apple iMac. When it was introduced in 1998, many predicted the end of Apple because the iMac lacked a floppy drive. At that time, users were still working with 3.5-inch floppy disks that fit 1.44 megabytes. If Steve Jobs' team had asked the users, they probably would have wanted bigger floppy drives (self-burning CDs already existed, but they were still extremely expensive, both the blanks and the CD burners themselves). Instead, Apple had built a modem into the computer; after all, it could also be used to exchange data.

An example from the Data Science world: A car manufacturer wants personalization on its website, just as other companies already have. In fact, this is difficult to implement, because for Spotify, it is easy to personalize song recommendations as enough data is available. That's not necessarily the case for visitors to a car manufacturer's website (even if slides from business consultants say otherwise). Can data still help make the site more relevant? Absolutely, though not in the way that marketers have imagined.

But even Data Scientists can quickly get lost in something that doesn't create value.[1] Just as you can lose yourself and your goal in an exploratory data analysis, it works the same way for many other Data Science activities. Trying out a new algorithm? A new development environment? Of course, time must be available for this as well, but quickly a system is created that is academically very interesting, but contributes nothing at all to the solution of the actual question.

3.2.3   Overcoming Resistance: Who Is Afraid of the Evil AI?

The use of AI, or even just machine learning models is not only met with jubilation. People fear for their jobs if a machine is to automatically evaluate data in the future and make decisions based on it. Existential fears arise here that cannot be soothed with "corporate blah-blah." The media fuel these fears by outlining dystopias in which even highly qualified employees no longer stand a chance against artificial intelligence. At the same time, reports like the one that a company like Zalando is laying off marketing staff and artificial intelligence is taking over have naturally caused anxiety. Even if many studies by management consultants on how many people will lose their jobs in the future are only written because they are intended to attract the attention of CEOs, there is no denying that there will be changes.

The emotional resistance that this creates should not be ignored under any circumstances. It is important that employees are also taken along the journey to automation through AI. More rationally oriented people may find this annoying, but if this step is omitted, resistance is to be expected, which can end in sabotage of the new system. And then, nothing would have been achieved, except losing a lot of time.

But, how do you do that? The *House of Change* by Claes F. Janssen, based on Elisabeth Kübler-Ross's grief curve, outlines the phases that employees go through:

1. 1st room, satisfaction: everything should remain as it is.

2. 2nd room, rejection: this change is bad, it can not work.

3. 3rd room, confusion: I can't cope with this situation; here also comes the rational and possibly also emotional insight.

4. 4th room, acceptance: I learn something new, and only now I understand why this is good.

---

[1] I don't refer to any Data Scientist I know personally, but I confess that this happened to me as well.

Employees cannot be expected to get from Room 1 to Room 4 right away; they usually have to go through Rooms 2 and 3 as well. They can be supported in this, but they will take their own time to do so. It is usually not the Data Scientist who leads this transformation task, but he needs to understand it. Since he is part of the renewal himself, he cannot always adequately understand the problems of the employees. Management is challenged here to find suitable measures to shape the transformation efforts in such a way that as many employees as possible join the transformation journey.

3.3   Basic Approaches in Machine Learning

This section is about acquiring the basic framework for machine learning tasks.

3.3.1   Supervised, Unsupervised, and Reinforcement Learning

Let's imagine a customer database that contains information, such as age, gender, previous purchases, and address, that is necessary for business operations. Many companies claim to have a lot of data but don't use it, and most of the time this is related to not asking the right questions of the data. What could be questions to the data in the customer database?

A first question might be whether there are groups within the clientele that are so far not obvious. The goal is to better understand the customer groups, but there is no target variable that an algorithm could learn from. This is called unsupervised learning. The algorithm tries to figure out which groups can be formed without any guidance. We will look at such algorithms in Chapter 7, "Clustering".

Another question might be which customers are more likely to cancel their contract, for example. Here, there is a target variable, namely, termination or no termination. In this case, historical data is looked at for customers who have cancelled and customers who have not cancelled, and these data are labeled accordingly, for example, 1 for cancellation and 0 for no cancellation. The label then contains the target variable. This approach falls under supervised learning and is discussed in Chapter 8, "Classification." The algorithm is supervised, that is, it is given what previous cases looked like with the labels so that it can apply the knowledge it has acquired to new cases. In the case of clustering (Chapter 7), such supervision is missing, and the algorithm is left on its own. Well, not entirely, because something has already been done by the human here by selecting suitable variables.

In addition to these two machine learning paradigms, there is a third one, reinforcement learning. Here, both worlds are combined to a certain extent, because the algorithm develops its own strategy and receives feedback, a reward, so to speak, so that the algorithm tries to increase the number of rewards. Often, the term "agent" is used in this context, and in fact, you can think of it as a virtual entity that develops its own strategy to solve a task based on praise and blame.

3.3.2 Feature Engineering

Feature Engineering is the selection and preparation of variables or features that have a high information content for the problem and are later fed into a machine learning model. Example: We want to predict whether a loan can be repaid by a potential borrower. The following features are available:

- Amount of the credit

- Credit history to date

- Zodiac sign of the borrower

- Account balance

Obviously, some of these variables are very interesting while the zodiac sign will contribute very little. This is an obvious example, but in reality, it is not always that clear. We will get to know several data sets in the next chapters, and again, we will examine the variables to see whether or not they help us solve a question.

3.4 Performance Measurement

After the features and parameters of a model are chosen, the model is applied and the performance of the model is measured.

3.4.1 Test and Training Data

To test a model, not all data are used for training (the so-called training data), the other part is used for testing (test data, also called holdout data). The results of the model for the test data are compared with the actual result. Under no circumstances should the training data also be used for testing. In unsupervised learning no training data is needed, since we have no target variables for the algorithm to learn from.

A special form of testing is the so-called *cross validation*. It is visualized in Figure 3.3. In the first iteration, one part of the data is used as a training set, and the other part as a test set. After that, the data is re-split, and two other subsets are used as training set and as test set. This is repeated several times;

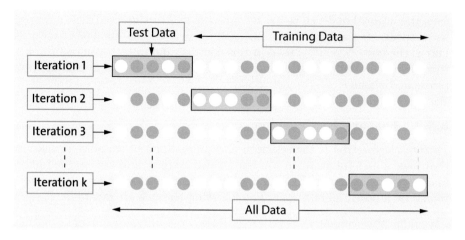

FIGURE 3.3
A k-fold cross validation.

in the figure, k stands for the number of iterations. However, no validation is as good as one with new, fresh data.

In the literature, a distinction is sometimes made between test data and validation data, but the terms are also often used interchangeably, which can lead to confusion. In fact, however, there is a good reason to distinguish them. Assuming that a model does not deliver good results at the beginning based on the test data, then the model is optimized based on these results (for example, by adjusting parameters). However, this then also means that this optimization works exactly for this test data. Therefore, it is often suggested that the training data is again divided into two parts, training data and validation data. The actual test set, however, which we separated at the beginning, is kept out of this process until the end, when a model is finally optimized. Only then is this test set used to test the final optimized model and to rule out the possibility that the optimization only works well for the validation data.

In this book we will mainly work with the simple form of validation and use only one holdout set. An example of cross validation can be found on the book's website.

3.4.2   Not all Errors Are Created Equal: False Positives and False Negatives

It is unlikely that a machine learning model will deliver error-free results. You have to expect errors. But, not all errors are equally bad.

In statistics a distinction is made between an *Type I Error* and *Type II Error*, also called alpha error and beta error. For example, a type I error occurs when a non-pregnant woman is told that she is pregnant although she is not, due to a false positive. The pregnancy test is positive when in fact, the woman is not pregnant. In contrast, a type II error occurs when the pregnancy test is negative, but the woman is pregnant (false negative). One may argue in this case what is worse. If there is a strong desire to have a child, then the disappointment of not being pregnant despite a positive test can be emotionally challenging. If a pregnant woman behaves unhealthily because she had a negative pregnancy test, she is threatening the health of her unborn child.

Similarly, when a disease is diagnosed that is not actually present, it causes the supposedly seriously ill patient to take actions that he or she would not normally take. Conversely, a seriously ill patient might have a different focus in their life if they knew about their illness. Another example is a fire alarm: If the alarm goes off even though there is no fire, it is annoying, but not so bad overall. If no alarm goes off even though there is a fire, then that can have disastrous consequences.

The two types of error also have an impact in business. If a machine learning system assumes that a loan application would result in a default even though the applicant is solvent enough to service the loan, then the credit house loses business. If a loan is approved even though the applicant will not be able to service it, the credit house suffers a loss. While losses are always to be expected, they can only be accepted by ensuring that enough business also takes place. So if an algorithm is too conservative and rejects too many applications, the credit house will have fewer losses, but at the same time, less business. If the credit house has too much risk appetite, then it will do more business, but will also have more losses. The losses could then eat up the profits from more business, because if only 1% interest is paid on a loan for an average loan volume of 10,000 dollars, then 100 such loans are needed to compensate for a loan loss of this amount!

It is the task of the project team, together with the business, to determine which error is more acceptable. Different algorithms can also have different tendencies in one direction or the other, and also, parameters in the modeling can be modified to influence the result. Also, a threshold can be set differently, for example when it comes to using a probability forecast for prediction. Staying with the credit institution example, if the threshold for rejecting a loan application is a 70% probability of default, how much damage would that mean and how much profit? Other interesting questions arise, such as what role the size of a loan plays. There may already exist in the model an influence of the loan amount on the probability of whether the loan can be serviced.

But, it would be fatal if the machine learning model were only wrong when it comes to high sums but always right on the low credits. Again, a thorough analysis is needed to test the impact of the model. Every company will want to try to minimize both errors and may even ask that the model optimizes both errors to 0. However, this is usually not possible, because the false positive rate and the false negative rate can be negatively correlated.

3.4.3 Confusion Matrix

A *confusion matrix*, also known as a *decision table* or *truth matrix*, is used for the evaluation of a binary classifier. Such a classifier decides whether something belongs to one class or another: pregnant or not pregnant, creditworthy or not creditworthy, spam email or legitimate email. Classifiers and confusion matrices with more than two classes also exist. An example with two classes can be seen in table 3.1.

In this context, two further concepts and their terms will be introduced: *specificity* and *sensitivity*. A classifier does not only make wrong decisions, which were the subject of the previous section, but also correct ones. In the confusion matrix, the true positives and the true negatives are also listed. The sensitivity indicates the probability with which a classifier correctly classifies a positive event as positive; therefore, it is sometimes also called *true positive rate*. Specificity indicates the probability with which a classifier correctly classifies a negative event as negative; hence the alternative term *true negative rate*. By looking at the table, it is possible to see how well a model performs in each class.

Sensitivity and specificity are not only known in machine learning; they are also used in medical tests, for example. If a COVID rapid test achieves a sensitivity of 99.76% and a specificity of 99.95%, this means that the test has detected 99.76% of the patients who actually have COVID and that it has also detected an actual healthy person in more than 99.95%. In a confusion matrix, it looks like table 3.2.

Obviously, one would wish here that the sensitivity would be better, that is, that the number of false negatives would be lower. One "disadvantage" of the confusion matrix is that it does not provide a single value with which to compare performance between different models. At the same time, this is also its advantage, as it makes it easier to understand the strengths and weaknesses of a model, especially when it comes to stakeholder engagement.

3.4.4 ROC AUC

The acronym ROC stands for receiver operating characteristics and dates back to a time when machine learning was not even in its infancy and, strictly speaking, the term had nothing to do with it. During the Second World War, the ROC curve was used for the analysis of radar signals. It was not until later that the concept of receiver operating characteristics made its way into ma-

	Loan repaid	Credit defaulted
Actually repaid	2,839	232
Actually failed	23	3,212

TABLE 3.1
Confusion Matrix.

	Test positive	Test negative
Actually positive	9,976	24
Actually negative	5	9,995

TABLE 3.2
Confusion Matrix for a Covid Test.

chine learning. The acronym AUC stands for area under the curve and refers to exactly that: the area underneath the Receiver Operator Characteristics graph.

An example of a ROC AUC graph is shown in Figure 3.4. We see two familiar features from the previous section: specificity and sensitivity. At the same time, it should be noticed that the scales of the axes differ: That of specificity starts at 1 on the left and ends at 0 on the right, whereas that of sensitivity starts at 0 on the bottom left and ends with 1 on the top. If the specificity is at 1, then the sensitivity is at 0, which is logical because if all cases are classified as negative, then all negative cases will be caught in any case. It just has the side effect that at the same time all cases that are actually positive are now also classified as negative. Conversely, if all cases are simply classified as positive, then the sensitivity is 1, because finally all positive cases have been

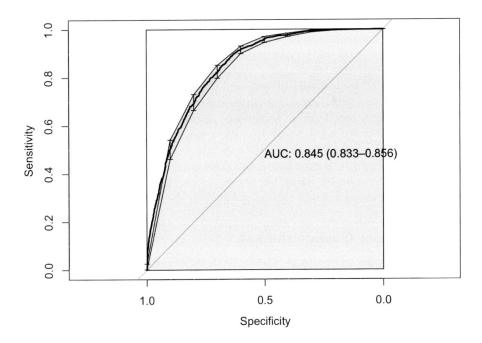

FIGURE 3.4
A ROC AUC graph.

correctly classified. Here, the side effect is that all events that are actually negative have also been classified as positive, and thus the specificity is 0.

The more the graph is oriented towards the upper left corner, the better, because eventually, it gets close to 1 for both dimensions, sensitivity and specificity. This makes the area under the graph larger, and that is exactly what the area under the curve is. In an ideal model, the line of the graph would go straight up the x-axis at 1 and then continue straight parallel to the x-axis at 1 on the y-axis.

The straight line in the graph connecting the points (0-1) and (1-0) means an ROC AUC of 0.5 and represents a classifier that could also discriminate simply at random. It would decide correctly in about half of the cases and wrongly in the other half.

The advantage of the ROC AUC is that you now have a single number that can be used to compare the performance of the model with that of other models. However, the number alone is still not enough, because the graph itself contains further information as displayed in Figure 3.4. You can see from the curve where the model is better and where it is worse.

3.4.5   Precision Recall Curve

The terms *precision* and *recall* are mainly known in the field of information retrieval, where they are used for the evaluation of search engines, for example. In very simplified terms, recall is the number of relevant documents found in relation to all documents in the index that are relevant. Precision is a measure of the accuracy of the ranking (are the most relevant documents on top?). In machine learning, recall is the same as sensitivity. Precision is the ratio of true positives divided by the sum of true and negative positives. You can see an example of such a precision-recall curve in Figure 3.5.

The precision recall curve, sometimes also referred to as the PR Curve, is helpful, especially when there is a *class imbalance* (see Chapter 5, "Exploratory Data Analysis"). This is the case when there are few cases in a class, for example, few proven insurance fraud cases compared with many legitimate insurance claim reports. Using the curve, we can accurately understand the behavior for a class.

3.4.6   Impact Outside the Lab

A model can have an enormously good ROC AUC or be at the top of another performance metric, but this does not mean that the model will actually perform well in real life. Since training is done with historical data, overfitting can be a problem. Very rarely is the performance of a model as good in real life as it is in the lab. But, changing conditions that lead to different data can also have an impact. The Corona pandemic is unfortunately a good example of this. These Data Science metrics are not exciting to stakeholders; they are not

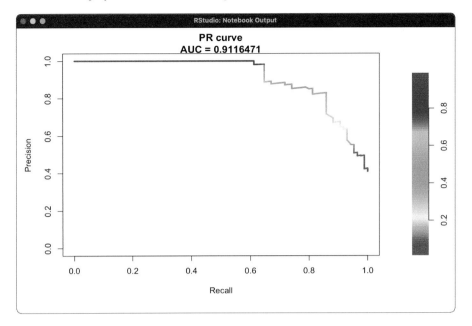

FIGURE 3.5
A precision recall curve.

interested in the ROC AUC. Instead, there are other metrics for the business or organization, for example:

- Sales

- Number of new customers that can be acquired

- Number of customers who did not terminate their contract after all

- Speed at which money can be collected

- Number of loans that are no longer in default

- Number of melanomas detected

The ROC AUC, or whatever other metric is used, should ideally correlate with the key performance indicators (KPIs) of the business. The better the model, the more business or treatment success should also be measurable. To test the effect of a model, *A/B tests* or *split tests* are usually used. This means that one part of the customers or users or patients experiences the previous application, while the other part experiences the results of the system. In this way, we can check whether the results of the model are statistically significantly better.

3.4.7 Data Science ROI

A Data Scientist costs money, the infrastructure to run a Data Science project costs money, and all the entities that support the Data Science project will consume money. The result of the project will go into production one day, there will be costs also then, for running the machines and for continuing to maintain the model. Data Science is not a cheap business. A Data Science project must therefore not be viewed solely from the perspective of model quality, but must be set in relation to costs. This is the return on investment (ROI).

3.5 Communication with Stakeholders

What is a stakeholder? According to the definition of the Project Management Body of Knowledge (PMBOK), a stakeholder is an individual, group, or organization that affects or is affected by, or believes it will be affected by [...] a project, program, or portfolio. Experience shows that never do all stakeholders feel sufficiently informed. The golden rule is: You can never communicate enough.

3.5.1 Reporting

Regular reporting is an important building block for the success of a project. On the one hand, regular reporting means that no stakeholder can say afterwards that they didn't know anything, and on the other hand, it also forces you to consider again and again what "business value" you have actually created with the project so far. The frequency depends on the project, but monthly reporting should be the minimum. In such a report, it is important to provide concrete figures. So, not "We received good feedback" (that reads to managers like "We had a nice chat over a cup of coffee"), but rather, "Seven of the nine data sources have been connected". Beware of percentages like "We are 80% done" because often the last 20% takes as long as the first 80% took. Ideally, a report should contain the following points, freely inspired by Scrum:

- What has been achieved (in numbers)?

- What's next?

- What are blockers?

In Chapter 5, I will discuss visualizations that are great for informing stakeholders about the progress of a Data Science project, but also for introducing them to details.

3.5.2 Storytelling

Even if it's not one of some people's personality preferences, communication is everything in a Data Science project. Some malicious tongues claim that it is more important for a Data Scientist to be able to create good PowerPoint slides than to generate great analyses and models. That's an exaggeration, of course, but there's also no denying that a bit of selling is also part of a Data Scientist's profile.

Storytelling can take different forms. For example, patterns of known stories can be used, which has the great advantage that people immediately listen better, because they already know the basic form of such a story and are eager to hear whether it also has the known ending. Example: You had a lot of trouble finding a feature that helps with classification. All previous attempts were crowned with little success. Now, they have identified a feature that has received little attention so far, but that has made a huge difference. That definitely sounds like Cinderella! Not only does it give you a good wrap for the previous unsuccessful efforts, but at the same time, you show that it wasn't easy to find this feature. But, you don't have to use a fairy tale right away, either. A story can also work on the basis of the following components:

- Setting the stage: What is the problem? How big is it? Why should it be solved?

- What has already been tried? Why was it not successful?

- What is being done now? What are the success figures so far?

- What can get in the way now?

- When can we expect the happy ending?

You should take into account that the recipients are usually not as deeply involved in a project as you are. They need more context ("What was that project again? Why are we doing this?"), which can be annoying for the data scientist himself ("I've explained everything before"). Here, it is necessary to understand that your own project is not the center of the world. Avoid technical terms that might be incomprehensible. For example, it is enough to talk about a model and results. The fact that XG Boost was used and the values of the confusion matrix look great can normally be kept to yourself.

3.6 From the Lab to the World: Data Science Applications in Production

Once a suitable model has been created, it should usually be put into productive use. The golden rule is to think about the end right at the beginning to ensure a project success.

3.6.1 Data Pipelines and Data Lakes

The data for a Data Science application is rarely, if ever, entered by a user in real time. Data can be rather static or change infrequently, such as historical data from order transactions, but it can also be absolutely *fresh data*, for example, data that flows in via a website or a customer database. Such data can, for example, be stored in a data lake from where it is processed using ETL-processes (Extract, Transform and Load, usually from different systems) via a data pipeline. to the target system. Setting up such architectures is usually not part of a data scientist's job, but it must be planned from the beginning.

Another point to consider right from the start is how to determine that a model needs to be updated. This can be done via monitoring, but also via checks at defined intervals. At the same time, you should define right from the start which KPIs are to be used to determine the need for an update.

3.6.2 Integration with other Systems

Often, a machine learning system is to be integrated into another system, for example via an API (application programming interface). This allows data to flow into a model and the results to come back out of the model into another system. Older systems often do not have the APIs that are necessary, and therefore, require costly customization. Verifying that the necessary APIs are in place is therefore part of the early project phase. If an API can be made available at a later stage, it must be continuously checked that this API also meets expectations. Nothing is worse than a Data Science project that delivers great results but is of no use to the user because it cannot be integrated.

3.7 Typical Roles in a Data Science Project

Data Science does not only consist of developing models, but usually also requires expertise in engineering and, above all, application knowledge for the area in which a Data Science project is to be carried out. However, small teams often do not have the luxury of hiring the appropriate experts for all tasks, so time must also be planned for the other tasks, and expertise must be built up.

3.7.1 Data Scientist

As is clear from the various definitions of Data Science, there are probably only a few people on this planet who can fulfill all the requirements placed on an ideal Data Scientist:

- An advanced knowledge of statistics and how to visualize data effectively

- Proficiency in multiple programming languages relevant to Data Science.

- Business knowledge

- Very good communication and presentation skills

It is more likely that a Data Scientist will be good at some of these areas, but not all, and that team members will complement each other.

How does one actually become a Data Scientist? Some universities offer individual courses, others have even set up entire degree programs. However, most Data Scientists are career changers, although that is probably not the right word, because since the profession of "Data Scientist" did not exist until a few years ago, Data Scientists without a degree of the same name tend to be those who have shaped the job description of a Data Scientist through their knowledge and expertise from other domains. An education in statistics, computer science, mathematics, or physics is definitely useful, but other courses of study can also enable one to find an approach to Data Science.

But, what exactly a Data Scientist needs to be able to do also depends on the environment in which she or he works. Small teams in particular do not have enough resources to employ a data engineer or a data science architect full-time. In such cases, this must also be handled by the Data Scientist. Quite apart from this, not all methods in the Data Science area are already so mature that they can be handed over to other areas in a sufficiently formalized way, as [42] put it:

> We must point out that Data Science, like computer science, is a young field. The par- ticular concerns of Data Science are fairly new and general principles are just beginning to emerge. The state of Data Science may be likened to that of chemistry in the mid-19th century, when theories and general principles were being formulated and the field was largely experi- mental. Every good chemist had to be a competent lab technician. Simi- larly, it is hard to imagine a working data scientist who is not proficient with certain sorts of software tools.

Even though this statement was made several years ago, it is still true today, although some principles have already developed.

3.7.2 Data Engineer

The occupational field of the data engineer is still relatively young and arises from the quote from the previous section. Just as a chemist in the early days had to be a good lab technician, data engineers are figuratively the lab technicians in Data Science. They provide data pipelines, for example, so that data from a data warehouse can be made available for a model. This data may also have different structures and is then brought into a uniform format by the data engineer. Likewise, the tasks of a data engineer include monitoring the applications created and getting them up and running again in the event of failures.

3.7.3 Data Science Architect

Just like the data engineer, the data science architect is is a new job description. They design architectures on cloud services such as AWS, Microsoft Azure or the Google Cloud Platform, but at the same time keep an eye on business requirements in terms of costs and time constraints.

3.7.4 Business Intelligence Analyst

A business intelligence analyst is often confused with a Data Scientist, as he or she already has advanced skills that enable competent handling of a data warehouse. A business intelligence analyst can and should also draw insights from data to identify opportunities for improvement within a company, but usually not with Data Science models. Although the lines between them and data scientist are blurred, business intelligence analysts are closer to the business, more likely to present their findings, and more likely to use Tableau and other business intelligence (BI) tools for visualization. They tend to do less programming than Data Scientists, and predictive models are more the province of the Data Scientist than the business intelligence analyst. But, this can vary from company to company.

3.7.5 The Subject Matter Expert

As a rule, it is necessary for the Data Scientist to familiarize himself with the context of a task. It is the subject matter expert who provides the Data Scientist with all the necessary information. Ideally, the Data Scientist even sits next to the employees working on tasks related to the system to be created. At the same time, the Data Scientist can involve the Ssbject matter expert early on to prototype and get feedback on whether the context is well understood. PowerPoint slides are one thing, existing software is another.

That's not always welcome. Most of the time, colleagues already have enough work on their plate, and they are not necessarily used to being heavily involved in development. There may be a prevailing expectation that requirements will simply be thrown over the fence, and the development team will come back later with the finished result. Not to mention that this type of software development is outdated for many areas, it will not benefit the end result. Instead, a commitment should be obtained right from the start that subject matter experts can be sufficiently involved.

3.7.6 Project Management

Even if some Data Scientists do not believe in the necessity of project management, or even believe that project management is not possible for Data Science, this is more likely due to a lack of knowledge about project management, so that the possibilities here are misjudged. In some Data Science books, project management is not even mentioned, for example in Papp et al. 2019 [40].

A frequently heard objection is that in Data Science projects, you can't predict how long something will take, because everything you're working on is new. In fact, however, this is exactly what projects are all about, as the sixth edition of the Project Management Body Of Knowledge Guide (PMBOK) of the Project Management Institute (PMI) from 2017 defines it [41]:

A project is a time-limited endeavor with the goal of creating a unique product, service, or result.

Thus, the PMBOK also mentions the development of a new pharmaceutical preparation as an example of a project, and past experience shows that not every such project is crowned with success. The project manager has several tasks within a project. In the classical view of the PMBOK she or he is responsible for "the achievement of the project's objectives".

- The project manager does not have to be the direct supervisor of the team members, but is responsible for what the team produces.

- Project management responsibilities include creating and monitoring adherence to schedules and budgets.

- Scope management.

- The project manager manages the communication with the sponsor of the project as well as with other stakeholders and the team members. Most importantly, there is the coordination of different and sometimes conflicting stakeholder goals that a project manager must balance.

- The project manager does not have to be a Data Scientist himself or herself. In the PMBOK, the role is compared to conducting an orchestra, which does not require mastering every instrument, but does require sufficient knowledge, understanding, and experience in the subject matter.

- Projects often compete for resources. In this context, the project manager must ensure that the project for which he or she is responsible has sufficient resources available.

- Determine and verify compliance with quality criteria.

- Procurement, often simply referred to as "purchasing."

Even if the role of the project manager is sometimes misunderstood as "doing everything the others don't want to or can't do," good project management is an essential factor for the success of a Data Science project. Or, to use another proverb from project management: If you fail to plan, you plan to fail. The classic project management process defined in the PMBOK provides for the following steps:

- Initiation

- Planning

- Execution

- Monitoring and control

- Conclusion

These steps don't look all that different from those of a Data Science project, which I presented in more detail at the beginning of this chapter:

- Business understanding

- Data understanding

- Data preparation

- Modeling

- Evaluation

- Deployment

This does not mean that the steps can proceed without iterations; just as the Project Management Institute has opened up to agile methods, Data Science projects cannot be planned using only the waterfall model.

3.7.7 Citizen Data Scientist

The role of the Citizen Data Scientist is a new development and also not yet set. A Citizen Data Scientist is not (yet) a real Data Scientist, but possesses knowledge that is suitable for more than simple data analysis activities. In fact, however, even perfectly trained Data Scientists are not always needed for projects. Among the activities mentioned in section 3.1, there are enough that do not necessarily have to be performed by a Data Scientist. These include, among others:

- Data collection

- Data cleaning

- Exploratory data analysis, where the focus is on "actionable insights"

- The creation of dashboards and reports

Citizen Data Scientists bring a very big advantage to companies: Instead of having a Data Scientist on every project that involves data, a Citizen Data Scientist can do the groundwork first and then bring in a Data Scientist when more complex problems need to be addressed. Often, Citizen Data Scientists are also subject matter experts, so there is no loss of information when transferring business requirements into the Data Science vocabulary.

3.7.8 Other Roles

A security advisor supports a project team in meeting the security requirements necessary for a project. This can concern the connection to APIs, the security of the network where the Data Science servers are located, but also the data flows within an application. Data leaks happen every week, and no one wants their employer or company to be in the newspaper because data was suddenly downloadable. For smaller companies that do not have their own security advisor, it is at least advisable to book a consultant who can take on this task.

Particularly in larger companies, the legal department also plays an important role in Data Science projects, for example when it comes to deciding which data may be processed at all. As a rule, lawyers are not trained in Data Science topics and need additional information to arrive at an assessment. In addition, there is still no case law at all on many topics within Data Science issues. It is naturally difficult for lawyers to arrive at an assessment, because what are they supposed to orient themselves by? Often, in-house counsel's job is to protect the company from harm, and so it is sometimes better to scrutinize new and difficult-to-assess issues, or at least certify them as a risk. This is a potential source of frustration for both sides, while both sides are merely performing their duties.

In some companies, there is a data steward, who has the task of ensuring the quality of the data and its sources. The data steward ensures that the strategic orientation of the so-called data governance is implemented correctly. Data governance refers to the processes, rules, and standards within an organization. A very simple example: Who is actually allowed to change data in a customer relationship management system when data is obviously wrong? The customer service representatives? The customers themselves? Assuming that a name change has occurred due to marriage, does this require a certificate to be presented as part of a know-your-customer-process? Who verifies this document? How is that documented? This is about a single data point, but every company has many data points that are handled differently. In Chapter 11, "Ethical Handling of Data and Algorithms," you will learn more about data protection, but surely (hopefully!), everyone is aware that certain data must be specially protected, and, for example, not every employee should have access to it.

The previous sections have described the variety of different tasks within the Data Science world, and it cannot be ruled out that additional fields of activity will be added.

4

Introduction to R

"People who love statistics are damned to hell for all eternity, people who like R even more so." [24]

R is a programming language and at the same time an environment for statistical data processing. Besides its specialization in statistics, R is famous for the possibilities of generating graphics to visualize data. R belongs to the GNU project and is available as free software. The programming paradigm of R, i.e. the principles underlying the design of the programming language, can be located in functional, object-oriented and dynamic programming. What this means exactly, I will explain in detail in the following sections.

4.1   R: Free, Portable, and Interactive

Programs that are written in R are not compiled, i.e., not compiled into a computer language. The instructions are executed in real time by an interpreter. Therefore, the code is usually easily portable to other systems, meaning that an R script written on the Mac will also run on a Windows system (but see the box below). In addition, you can simply type a command into the console and it will be executed immediately, without a more or less long compilation process.

The exception proves the rule.

If packages are used (see Section 4.1.2) that are not available
for all platforms, then a script is not easily portable.

In contrast to SPSS, which was a standard software for statistics in the 1970s to 1990s of and has belonged to IBM since 2009, R is a completely free solution.[1] However, it currently offers only some of the functionality that

---

[1] To be fair, we should mention that there is also an open-source variant of SPSS called PSPP.

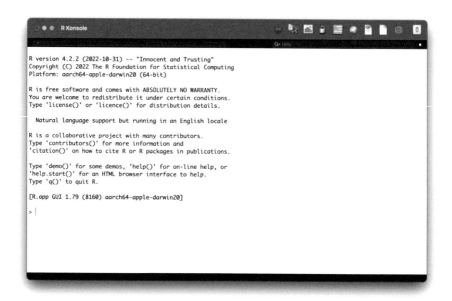

FIGURE 4.1
The basic R window.

SPSS does. In contrast to SPSS, R looks rather "naked" at first, because it does not come with a graphical user interface like SPSS where you can move to the result with inputs in masks and clicks through menus. Instead, the R environment initially provides only a command prompt (see Figure 4.1), which reminds the older ones of MS-DOS and confronts the younger ones with the question of what to enter.

SPSS also started as a programming language (initially SPSS stood for Statistical Package for the Social Sciences, later for Superior Performing Software System). Programmers still had to enter their programs and data with punch cards. Today, it is still possible to write programs in a separate programming language within SPSS. Meanwhile, R is becoming more and more the standard for statistical problems, not only in statistics, but in science in general, and also in business. This is not only due to the pressure to save money at universities and in the economy (where non-commercial software tends to meet with little approval if there is no commercial support for it), but mainly because R offers the user complete control over the analysis.

R is available for many platforms. Since it is not programmed in Java (like the latest versions of SPSS, for example), R does not require a virtual machine.

4.1.1 History

R is an open-source implementation of the commercial statistical programming language *S* and was released in 1993. Statisticians Ross Ihaka and Robert Gentleman began developing R in 1991 as an alternative to S, and used the first letter of their first names as the name for this new language. At the same time, they wanted to make the relationship to the language S clear. In R, references to S are found again and again, especially when R experts talk about S3 and S4 objects, describing the compatibility with versions of the language S.

Shortly after its release, a community emerged to provide feedback to developers. In 1997, the R Development Core Team was formed, which is still responsible for further development today. In the same year, the Comprehensive R Archive Network was born (CRAN) where documentation and packages are available for download.

In addition to the R Development Core Team, the R Foundation and the R Consortium take care of the language. The R Foundation is a non-profit organization based in Austria that manages the copyright and also the documentation of R. The R Consortium is funded by companies such as Google, Microsoft, Posit PBC, GSK and Merck. The mission of the consortium is to support the R Foundation by funding projects that facilitate access to the language and tools.

Since the year 2000, R has existed in version 1.0, as the development team assumed that this was the first stable version. In 2020, version 4.0 was released.

What do Charlie Brown and the Peanuts have to do with R?

In addition to the version number, R versions have names that sometimes come from the context of Charlie Brown and the Peanuts. Version 4.0.3, which is current as I write the first manuscript of this book, is called "Bunny-Wunnies Freak Out," based on the fictional book "The Six Bunny-Wunnies Freak Out." The book is banned from the school library in a 1972 comic strip, leading Linus, Sally, and Charlie Brown to launch a campaign against the book's banishment.

4.1.2 Extension with Packages

R offers many functions but as in almost any other programming language, R's functions can be extended with additional software *libraries*. These libraries are called *packages*. The choice of words in R is not always consistent, since packages are installed with *install.packages($'PACKAGE - NAME'$)* but later included with the *library($PACKAGE - NAME$)* command.

Before you install a package, you should check if there is not a way to do the project with the basic functions and the packages already included in it, also called Base R packages. Sometimes, packages simplify something that would also work with Base R, but would be incomparably more complicated. As an example, the Tidyverse, which will be the topic of Chapter 5, dramatically simplifies the data transformation process. Almost everything included in Tidyverse can also be solved somehow with Base R. It's just much easier with the Tidyverse in some cases. On the other hand, it is not necessary, for example, to create a graph with the package ggplot2, if it works with the simple plot command in Base R without any problems and the result looks just as good.

Many packages can be downloaded from the CRAN, as mentioned in the previous section. Other packages are available from other sources, for example GitHub. In December 2020, over 16,000 different packages existed on the CRAN server.

Whatever you can think of, someone will probably have already developed a library for it. Especially if something is needed again and again – then, it is usually worth to develop a library for it. Not all packages are good, even if they go through a hard process on CRAN before they are accepted. Therefore, it is worth having a closer look at the author of a package and his reputation.

Like R and RStudio, packages should also be updated, but not simply because a new version has been released. Updates do not happen automatically, but are triggered manually via TOOLS – CHECK FOR PACKAGE UPDATES. The version numbers of the installed packages are compared with those on the CRAN servers. Sometimes, these updates contain changes that cause existing scripts to stop running. This rarely happens, because development teams usually try to ensure backwards compatibility, but this is not always possible. Libraries used in production scripts, and programs that run in an automated fashion or in an application available to users, for example, should be examined beforehand for changes during updates. With a large number of packages installed by most R developers, this is not an easy task. Tt can be helpful to use a second environment for testing.

4.1.3   The IDE RStudio

RStudio is an integrated development environment (IDE) for R. Although RStudio is still more rudimentary in the interface than SPSS, it is already much more comfortable than the completely "naked" R. All important functions and programs for the development of software or the handling of data can be operated via a graphical user interface. There is no "media break", since one does not have to change in between to the file browser or to the terminal.

RStudio is developed by the company Posit PBC (formerly RStudio PBC) and was announced in 2011. In 2016, version 1.0 was released. RStudio cannot run without R. You must first install R, then RStudio. The latter is a kind of "add-on" to R. In Sections 4.2.2 to 4.2.5, I explain the installation, configuration and usage of RStudio in detail.

4.1.4 R versus Python

Python is the other big top dog on the terrain of Data Science programming languages. The first full version appeared in 1994, and contrary to what some followers believe, the name of the programming language has nothing to do with the snake, but with the British comedy group Monty Python.

There is some debate in the Data Science community as to which of the two is the better language to learn Data Science. The answer is simple: It is important to learn the basics of statistics first before thinking about the language. And even after that, it's not an either–or. First, the facts:

- R is open source (as is Python).

- It works on multiple platforms (as does Python).

- It is a statistics programming language (Python is not, but can be extended).

- Beautiful graphics can be produced with it (Python can do that too).

- R can be connected to other languages (Python can do that too, slightly more elaborately).

- It is flexible and extensible (as is Python).

Python is a powerful general-purpose programming language – as enthusiastic users like to point out – but not a language that focuses on statistics. Of course, there are many extensions for Python that add more statistical functions, but that's the point: they are extensions and not basic components like in R. In contrast, R includes almost everything in the standard installation that is needed for simple statistical work; it is just designed for that. But, R can also be extended. And, in addition, many libraries exist for R, which retrofit functions that Python possibly already masters with on-board means.

Both languages can be combined in RStudio (more on that later). Some say that it is a big advantage that R was developed by statisticians for statisticians, and the development team had exactly the needs of statisticians in mind. However, this may also be one of the disadvantages of the language.

When it comes to the popularity of programming languages, statistics are often cited from Stack Overflow, the question-and-answer platform for developers, in which the number of questions/answers is taken as a proxy for programming language popularity. This sounds plausible at first glance, but if you think about it more carefully, the number of questions could also be an indicator that users have more problems with a programming language.

The good news is that if you learned R as your first programming language, then learning Python will be very easy for you. And I can assure you that both programming languages can be a lot of fun.

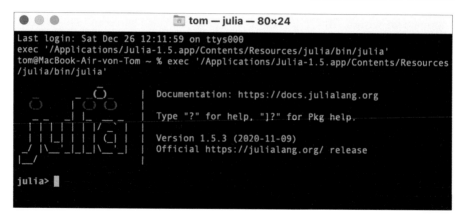

FIGURE 4.2
Julia in the macOS terminal looks something like R "pure".

4.1.5 Other Languages

I have already covered SPSS, other companions of R are for example SAS, Julia, and MATLAB. In this section, some principles of the respective languages are presented.

The name of the commercial statistical software SAS originally stood for Statistical Analysis System, but today it is simply referred to as SAS. It offers a development environment similar to RStudio. Users like to use SAS because no programming is necessary for simple analyses.

The mathematical programming language Julia was developed at the Massachusetts Institute of Technology (the famous MIT) and appeared in 2012. It has a high execution speed, but due to its young age, does not yet have the functionality of other languages. Julia is mainly designed for solving numerical problems. Like R, Julia can be used interactively in a console. More information can be found at https://julialang.org. Julia is especially interesting when speed is essential and cannot be satisfactorily guaranteed with other approaches.

MATLAB is, like SAS, a commercial software, published by MathWorks. Released in 1994, it is primarily designed for solving mathematical problems and visualization. Models can be trained and compared here via a graphical user interface. The special feature of MATLAB is that C/C++ code can be automatically generated from the models created. MATLAB is widely used in engineering, for example in mechanical engineering.

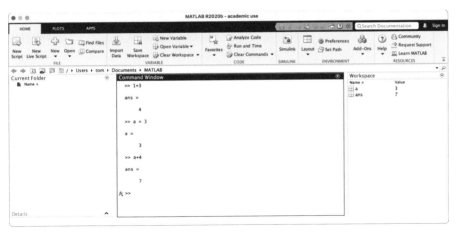

FIGURE 4.3
The MATLAB® interface. Similar to RStudio, there is an interactive console and a workspace.

4.2 Installation and Configuration of R and RStudio

This chapter is about the optimal installation and configuration of our tools. Even if you should have already installed R and RStudio, you will find tips here that can speed up your work with R.

4.2.1 Installation of R and Short Functional Test

You can download R from the official R-Project website at https://www.r-project.org. To do this, select the version for your operating system, and then, pick the mirror that is closest to your location.

Mirror server

Mirror servers may seem a bit old-fashioned, but they help to ensure that large data packets do not have to be transported halfway across the internet but instead, come from the nearest server. Theoretically, the software should then also be with you sooner.

After installation and startup, R greets you with a command prompt, or prompt, like a UNIX/Linux terminal or the venerable DOS user interface. The prompt tells you that R has nothing to do right now:

```
>
```

In this book, the prompt is always included in the listings so that you can distinguish the input of commands from the output. However, do not type the prompt with it.

You can use R in its simplest form by typing an expression into the console and then pressing Enter. The result then appears below the command you typed (see Figure 4.4). Some commands can also be executed with the menu, but analysts spend most of their time here using the so-called console when not writing a program.

R on the Mac

For macOS, there is an additional website of the R project at https://mac.r-project.org. It is interesting in this context that after the release of the Apple Silicon processors, R could only run with the Rosetta emulation for the time being, since the necessary FORTRAN compiler had not yet been ported to the new platform. Like many machine learning libraries, R depends heavily on FORTRAN. This is surprising in that FORTRAN is one of the oldest programming languages around. Developed in 1957 by IBM, its focus is on numerical computations, and it has thus made the leap from punch cards into the modern age. Today, an optimized version for Apple Silicon is available.

Try it out for yourself. Type (without the prompt)

```
> print("hello world")
```

and press Enter. The result will then appear below the command:

```
[1] "hello world"
>
```

You should have noticed that R has assisted you in typing: as soon as you type the first parenthesis, R adds the other part of the parenthesis to the right of the cursor. If you then enter a quotation mark, R adds that too. All you actually have to enter is:

```
> print("hello world
```

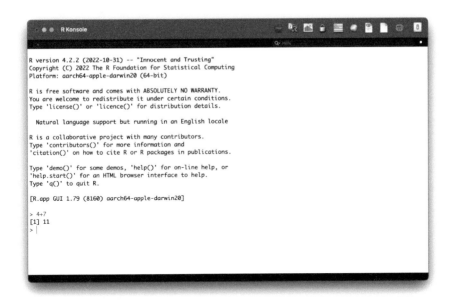

FIGURE 4.4
Running a command in R.

R should have done the rest for you. Programmers are lazy, and if there is a way to automate processes, they will be automated.

If you want to write a longer program, it makes little sense to enter the program code each time. Therefore programs are mostly written in files, for example hello_world.R, and these files are then executed in the R console after they have been loaded. R offers a very simple editor; it may be advisable to look for an alternative unless you want to switch to RStudio right away. Many well-known editors offer extensions for R, for example the venerable Emacs. The integrated development environment, RStudio, whose installation I cover in the next section, also includes a passable editor that I recommend for further use in this book.

4.2.2 RStudio Installation

The RStudio download package can be found on the Posit website at https://posit.com. RStudio is also free in the "community version"; additional paid licensing options are offered on the website. You can install RStudio either as a desktop application or on a server. In this book, I primarily describe the desktop variant. For installation, use the respective installer for your operating system.

FIGURE 4.5
The RStudio interface after the first startup.

After R and RStudio are installed, and RStudio is started (R only needs to be installed before RStudio, but does not need to be started before RStudio is started), RStudio appears with three or four windows, respectively, whose functions you can configure differently (see Figure 4.5). A detailed tour of RStudio can be found in Section 4.2.4. To check whether R also works properly in RStudio, enter a command in the console as you would with the "pure" version R.

4.2.3 Configuration of R and RStudio

You can now start right away, because for simple projects, the default settings are mostly suitable. But as soon as you start working with somewhat larger data sets, depending on the available memory and the complexity of the project, an adjustment of the configuration is necessary. This is due to a peculiarity of R, namely how R handles memory.

R loads all data into the main memory so that one quickly reaches the limits on a normal desktop computer. For example, if a computer has 8 GB of RAM, the operating system and all open programs already need part of it. If you still have 4 GB of the 8 GB left, it would therefore theoretically be possible to use a data set of 4 GB.

In reality, however, it's all a bit more complicated. This is because some UNIX-based operating systems, such as macOS, outsource part of the working memory to the hard disk (swapping), so that in principle the limit of 8 GB doesn't have to be quite so hard. This even happens when there is actually still enough physical working memory.

However, it is rarely the case that we work only with one data set that stays under this memory limit because additional memory is often needed for the data cleaning and transformation discussed in Chapter 3, "Anatomy of a Data Science Project". Unfortunately, operating system swapping is not as fast as using real RAM, since data must first be swapped out of RAM to disk and then read back into it. Hard disks are also usually not as fast as working memory, even if they are solid-state disks. Also, it cannot be said that swapping can work at all if there is not enough hard disk space left.

Back to R: When working in the R environment, memory is needed for various processes, for example, for each object that is created (I will discuss objects in more detail in the following sections). The environment in which you work during a session is also called a workspace. An object can be, among other things, a list that can contain a dozen or even several million values. This list is loaded completely into memory, whether it is read in from an external file or generated by a script during the session. If the list is then manipulated, for example duplicates are sorted out, then this also happens in the working memory. Since some operations are performed on a copy of the actual data set, for example, in order not to have to load all the data again and perform the previously performed operations again in the event of an error, it can quickly happen that different copies, more or less heavily modified, are located in the working memory. It is therefore advisable first of all to always clean up the workspace and delete objects that are no longer required. This can be done, for example, with the command rm(), which stands for remove as in UNIX, but with brackets containing the name of the object to be deleted.

How does swapping work?

Swapping is sometimes the reason why much less disk space is available than there actually is, because the operating system has swapped out working memory into swap files. Andrew S. Tanenbaum explains the principle of swapping, paging and virtual memory very well in his computer science standard work *Modern Operating Systems* [46]. Incidentally, Tanenbaum was also the one who said in 1992 in the Usenet group comp.us.minix that Linux, which had just been developed, was obsolete, which led to a flame war between Professor Tanenbaum and Linus Torvalds, then a student and Linux developer. Despite his misjudgment, Tanenbaum is still a luminary.

FIGURE 4.6
R exit dialogue.

R saves a workspace to the .RData file and loads it again when R is restarted. Although you will be asked when you exit R whether you want to save the workspace or not (Figure 4.6), the default action is to save it. This can cause problems when R or RStudio is restarted and a workspace or project is loaded that requires a lot of memory. R and RStudio should be configured so that this doesn't happen, because R and RStudio both have the nasty habit of just not responding if it takes a little longer to load. And sometimes, you don't know if something is still happening or not. In the worst case, R is busy, but the user doesn't see it and then quits the program via Ctrl+Alt+Del or some other kill command. R and RStudio can sometimes be a bit bitchy about response behavior.

I recommend not saving the workspace, even if R provides this as a standard process. Especially with larger data sets in the workspace, this can lead to problems when the workspace is read in again.

Unfortunately, in R, you cannot configure that R does not load the previous workspace on startup. It is only possible to disable asking afterwards and not save the workspace on exit.

In the settings for R shown in Figure 4.7 (MENU – SETTINGS – STARTUP SETTINGS), the loading of the history file at startup is deactivated, but it is activated that duplicates in the history are deleted during a session. The history file stores the commands entered so that they can be

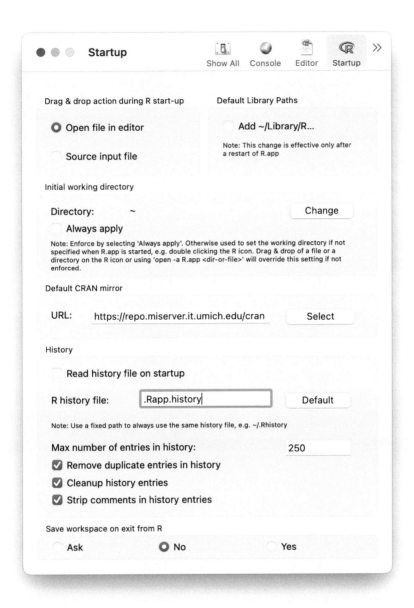

FIGURE 4.7
R settings

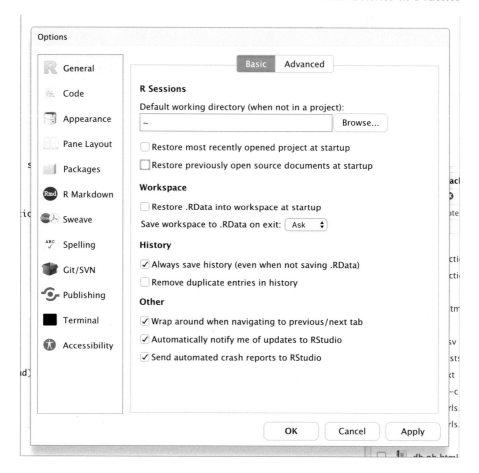

FIGURE 4.8
RStudio global options.

conveniently recalled. In addition, a preferred CRAN server is set from which packages are loaded. Just as with the mirror concept explained earlier, the idea is to avoid unnecessary load on the Internet by getting the software from the server closest to your own location.

In the global settings of RStudio, on the other hand, it is possible to disable loading of the workspace from .RData at startup. Additionally, in the example in Figure 4.8, the option is deactivated that the last loaded project as well as previously opened source files are loaded again. These settings also ensure that RStudio does not appear to simply refuse service because something is still loading in the background.

Some programmers want to customize their interface to their personal preferences. To do this, you can change not only the order of the windows in

RStudio, but also their appearance. For example, many prefer to use a dark background when programming so that the eyes do not tire too much.

4.2.4 A Tour of RStudio

We will now look at the individual areas of RStudio in more detail. A complete manual for RStudio would go beyond the scope of this book, so I will only explain the most important functions. If you open RStudio without having worked with it before, it should look like Figure 4.5. On the left side you see exactly what you see in the R application, the console. On the top right, you see your working environment with more tabs, and on the bottom right, your files and more tabs.

Now, enter:

```
> View(cars)
>
```

The capital *V* is important, because otherwise, you will get an error message. The console should have "slid" down, and RStudio looks like it does in Figure 4.9. In the upper left part (called SOURCE, even though it doesn't say that anywhere) you should see the *data frame* of the data set cars (we will cover data frames in a few paragraphs), which is supplied with R. The data set consists of two variables: the speed of cars in miles per hour and the

FIGURE 4.9
The four main areas of RStudio.

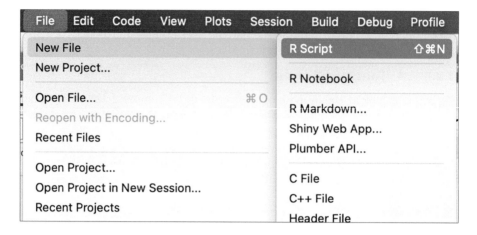

FIGURE 4.10
Creating a new file via the menu.

distance needed to brake. The data was collected in 1920. You can get more information by typing ?cars in the console.

Usually, you do not work with only one tab, but have many tabs open due to creating scripts and notebooks as well as opening several data sets. It is not uncommon, then, to have so many tabs open that what you are looking for is not immediately visible. Somewhat inconspicuous on the left under the tabs are arrows with which you can navigate through the tabs.

You can create new files in RStudio in two ways: either via the menu (see Figure 4.10) or via the small plus sign in the respective workspace (see Figure 4.11). As a first example, I will use the integrated editor.

Several different options are available in it; please choose the first option, R SCRIPT, for the first example, whether from the menu or from the workspace. Now, a fourth window should open inside RStudio, above the console.

A simple R script is created in this editor. Now, enter the code you used before:

```
print("hello world")
```

Again, R assists you in typing so that you only have to type part of the command. Click on the green arrow with the RUN command next to it at the top right of the window, and the code will be executed and the result printed in the lower console window (see Figure 4.12). Please leave the tab open, you will reuse it later.

Depending on which tab is open, different buttons are visible within the tab. Only two groups of buttons are common to both:

- Scrolling between previous actions. Left arrow goes back, right arrow goes forward.

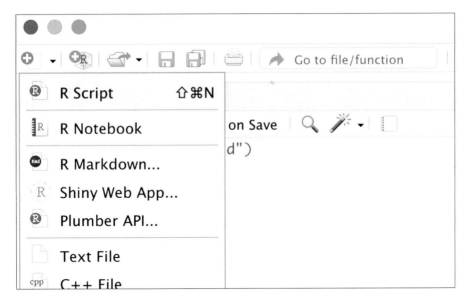

FIGURE 4.11
Creating a new file within the respective working environment.

- SHOW IN A new Window – Open the record or script in a new window. The tab will then be closed and will not be reopened when the additional window is closed again. In addition, you have the possibility to filter the data frame and to search in it. However, you cannot change data here! You can do this in R by entering edit(cars), but this command does not work in RStudio. If an editor is open, there are additional buttons for saving the code, search and replace, and, depending on the document type, other options. For a Markdown document (see Section 5.3.1, "Why use the Tidyverse?"), for example, there are:
  - a spell checker
  - a button for generating a preview for "knitted" documents as well as various knit options (see Section 5.3.2, "The basic verbs")
  - various document options If a simple R script is loaded, in addition to code tools at the magic wand icon, there is also the possibility to generate a report that saves the code as well as its output.
- In the lower left area you can see at least three tabs:
  - the console that offers exactly the interface as R without RStudio
  - a terminal window that offers different shells depending on the operating system
  - jobs in which processes that do not run via the console can be seen

FIGURE 4.12
The RStudio interface with an open script whose execution result is displayed at the bottom of the console.

There may be additional tabs, for example, for R Markdown.

In the upper right area, the first tab is the ENVIRONMENT, a part of the working environment that lists all objects loaded into memory. Packages that are also in memory are not listed here, so it is only a part of the working environment. If you have run view(cars), you should also see this data frame here. Close the data frame, and then click on the name of the data frame in the list so that it opens again. The Environment is an extremely useful feature of RStudio because otherwise, you would have to remember what is already in memory and what is not.

The second tab lists the history of the entered commands. The third tab shows connections, for example if you have established one to a Spark cluster (see Section 10.2.2). The fourth tab lists various Git-options, which I will discuss in Section 10.1. The last tab is for tutorials.

The area at the bottom right contains the following tabs:

- FILES: a kind of Windows Explorer for the computer's file system.

- PLOTS: plots generated via commands in the console or in the code editor are output here.

- PACKAGES: all installed packages; those that are loaded are highlighted. Here you can also load new packages.

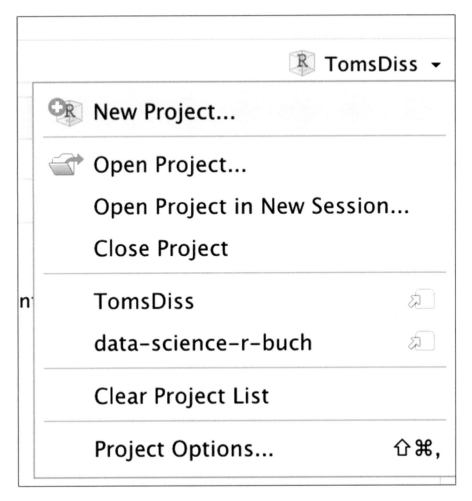

FIGURE 4.13
The project menu in RStudio, top right of the interface

- HELP: all help documents
- VIEWER: here, for example, local HTML documents generated by scripts are displayed.

4.2.5 Projects in RStudio

A special feature in RStudio is projects. If a project is created, the respective workspace is also loaded when switching from one project to the next, including the working directory. The projects can be found at the top right of the RStudio interface. If you decide to create a project, you can also choose from a variety of templates, for example Shiny apps, Markdown web pages, R packages and much more.

Projects and their advantages

I always recommend using a project unless you just want to try something out really quickly. Because a common problem that my students, for example, have to deal with is the working directories of a computer. Files are saved and can't be found later, or data should be read in, but you don't know how to specify a path under Windows (C:\Users\) or the Mac (/Users/NAME/), so you get error messages. You don't have all these problems with projects. Of course, it's also beneficial to get to know your own computer and how it works, but if you want to be a little more organized in your approach to your own projects, the projects in RStudio are recommended, especially in conjunction with GitLab or GitHub (see Chapter 10, "Workflows and Tools").

When you create a new project, RStudio offers you several options as shown in Figure 4.14. For your first attempts, you should simply create a fresh project, in a new directory. I will discuss "Version Control" in Chapter 10.

You should always open a new project in a new session so that the objects from another session do not collide with those of the new project.

4.2.6   The Cloud Alternative: Posit Cloud

RStudio also offers a cloud solution, Posit Cloud, previously RStudio Cloud (see figure 4.15), discounted powerful packages are available for students and teachers. A free version with a limited number of project hours is also on offer. As with the server package, you can work independently of your own computer.

Posit Cloud (posit.cloud) structures everything into projects. Similar to the desktop version, projects can be started with a version control repository.

The packages each contain a different number of project hours, which are composed of usage, number of CPUs, and memory. The good thing is that you can use different configurations depending on the project. However, the number of CPUs is still limited to four and the RAM to 16 GB in winter 2021/2022. If you need more, you should go for the server variant that runs on AWS or another cloud provider.

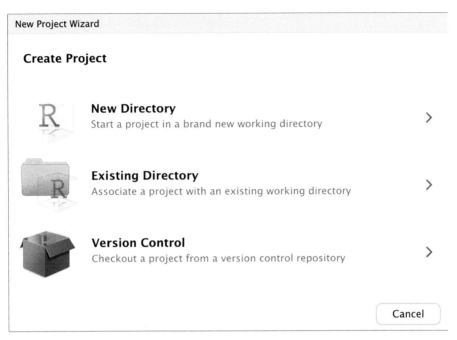

FIGURE 4.14
Different options when creating a new project.

4.3 First Steps with R

Now that we have installed and configured R and RStudio, we will start with the introduction to R. If you have programmed before, you will find many elements that are similar in most programming languages. Nevertheless, R sometimes has its own view on things, which you will learn in the very next section.

4.3.1 Everything in R Is an Object

In the R world, everything that exists is an object, and everything that happens is a function call. But in R, even functions are objects, and even the variables that reference an object are themselves an object and can be manipulated.

Understanding this, that everything is an object, is elementary to further understanding R. In the next sections, I will introduce you to the basics.

4.3.2 Basic Commands

You can use R like a calculator, as described earlier, by entering operations directly into the console:

FIGURE 4.15
Using Posit Cloud, R and RStudio easily in the browser.

```
> 7+4
[1] 11
>
```

The result of such an operation can be stored in a *variable*. A variable is one of the simplest objects in R. A variable is like a container in which a value can be stored. If you assign a new value to the variable, the old value can no longer be retrieved. In the following example, the result of an arithmetic operation is assigned to the variable result:

```
> result <- 7 + 4
>
```

The expression could be phrased colloquially as follows: Let result be the result of $7 + 4$. Incidentally, the arrow can also point in the other direction, but you should avoid doing so if possible.

The difference between $< -$ and $=$

For beginners, it can be confusing that there is both the arrow $< -$ in R and the equal sign $=$. Basically, you can remember that the arrow is used to create objects, as we just did, while the equal sign is used in functions. I will point this out to you again in the relevant places.

If the result of an operation is stored in a variable, the result cannot be seen immediately. The variable name must be entered again to see the result:

```
> result
[1] 11
>
```

However, you can also view the result immediately by placing the expression in parentheses:

```
> (result <- 7 + 4)
[1] 11
>
```

Upper and lower case must be observed:

```
> Result
Error: Object 'Result' not found
>
A space is not allowed in variable names:
> second result <- 7-2
Error: unexpected symbol in "second result
>
```

A hyphen in the result is also of little use, because it is interpreted as a minus:

```
> result-2 <- 3
Error in result - 2 <- 3 : could not find function "-<-".
> result-2
[1] 9
>
```

Special characters in variable names are possible, but your code will then not necessarily work on every system, which is disadvantageous if you want to pass it on to other users.

4.3.3 Data Types

A data type of an object determines what we can do with the object. In our examples so far, we have assigned only numbers to the variables. Of course, there are other data types, and numbers are also further differentiated:

- Character: any type of character or string of characters.

- Numeric (for example, 2 or 2.1, numeric)

- Integer (an integer, i.e. also 2, more about that in a moment, integer)

- Logical (true or false)

- Complex (not discussed further in the book)

- Raw (not discussed further in the book)

In the R language, a variable can take on a different type at any time.

```
> result <- 2 + 4
> result
[1] 6
> result <- "4"
> result
[1] "4"
>
```

Text, whether a single letter or a string, has the type character. Text must be enclosed in quotation marks. This *dynamic typing*, where the variable type does not have to be declared in advance and the variable can take on any type, even within a program, has advantages and disadvantages. The advantage is that you can quickly change something, and the disadvantage is that this "quickly change something" also quickly leads to errors.

With typeof() the R-internal type of an object can be displayed:

```
> my_variable <- 2
> typeof(my_variable)
[1] "double"
> my_variable <- "2"
> typeof(my_variable)
[1] "character"
> my_variable <- 2L
> typeof(my_variable)
[1] "integer"
```

Here, you can also see the solution for how a number can take the data type Integer: There must be an L after the number. At the same time, you see here for the first time the use of a function. You can recognize functions by the fact that they always have an opening and a closing parenthesis after the command, like typeof(). Inside the brackets, *arguments*, also called *parameters*, are listed. A function can expect one or more arguments, and there are also functions that do not require an argument, such as:

```
> date()
[1] "Thu Aug 26 21:04:56 2021"
```

The biggest challenge is often remembering which parameters a function requires. You will get to know more elementary functions on the next pages, which you will probably work with frequently.

A typical source of error is that the data type of an object is different than you think:

```
> result * 3
Error in result * 3 : non-numeric argument for binary operator
>
```

Since we had assigned a string to result, the variable can no longer be multiplied by a number. The data type can be changed at any time by as. , followed by the desired data type, provided the operation can be supported, for example:

```
as.charachter(result)
```

A date is a special data type and can take different forms: for example, May 1, 2019 or 01/05/2019. In the last case, however, it is not necessarily clear whether it is May 1 or January 5, because in the USA, the month is mentioned first, then the day. Such a date format also has the disadvantage that you have to convert it first to perform arithmetic operations with it, for example if you want to know how many days lie between May 1 and July 27.

One way to work with data more conveniently is the UNIX Timestamp that uses the number of seconds since 01/01/1970. This number can be converted into a real date relatively easily.

The year 2038 problem

You may remember the Y2K problem: Some programs had simply used the last two digits instead of a four-digit year, and that's what was supposed to be a problem when jumping to the year 2000. Surprise: The year 2038 will hold a similar challenge for us, because then some UNIX systems will no longer be able to continue counting, to put it simply, and will then jump back to 0.

The knowledge of data types is also important because R decides itself which data type a variable has when importing data without additional specification. Sometimes, this leads to problems, if the data cannot be evaluated correctly after the import.

R offers several data structures that are also known from other programming languages, although some of them work somewhat differently. Here are the most important data structures:

- Vector

- List

- Matrix

- data frame

- Factor

Vectors are the simplest object in R. They are a one-dimensional collection of data points. With the function c() (for concatenate or combine), you tell R to create a vector.

```
> x <- c(1,2,3,4)
> x
[1] 1 2 3 4
>
```

In the output, the number at the beginning of each line indicates the index value of the first value. Each value gets its own index number. Unlike many other programming languages, R starts counting at 1, not at 0.

As mentioned before, programmers are lazy people: they try to simplify everything that needs to be done more often, so that this vector can be created more easily as well:

```
x <- c(1:4)
```

Vectors can contain only data of the same type:

```
> x <- c(1,3, "4")
> x
[1] "1" "3" "4"
>
```

The numbers have suddenly become strings, just by the fact that the last number was put in quotation marks. This is a common source of errors, because if we don't have any numbers, although we actually expect them, then they cannot be used in calculations, as you have already seen above.

Lists are a data structure in R, which, unlike what one usually imagines a list to be, can consist not only of one row of values, but of several rows, which can also have different lengths.

```
> x <- c(1:4)
> y <- c("1", "2", "3", "4", "5")
> my_list <- list(x,y)
> my_list
[[1]]
[1] 1 2 3 4
[[2]]
 [1] "1" "2" "3" "4" "5"
```

If a single value is to be called, you must specify two index values:

```
> my_list[[2]][2]
[1] "2"
>
```

Data frames are an essential concept in R (and in Python, too, by the way). A data frame is like a kind of Excel table with observations in the rows and variables in the columns. Each column is a vector.

Rows or columns: Important!

The concept of data frames is so important that it should be explained in more detail. If you ask all the people you meet in the pedestrian zone for their height and weight, then a row is created in a data frame for each pedestrian with at least two columns, height and weight (and perhaps the pedestrian's consecutive count number in the first column). Each pedestrian has its own row and is a *statistical unit* or *observation*,; each column is a variable (which consists of a vector as described above) or a characteristic. In short: Observations in rows, variables in columns, this is how the ideal table for data analysis is structured. Often, this format is also called *Rectangular Data*. Sometimes this format is also called *Wide Format*.

Not all data is available in this form, and often, enough data must then be converted to this format so that it can be used for machine learning models. However, some models also require a different input.

Let's take a look at one of the data frames already provided in R. By entering the function head(), the first lines of the data frame are displayed:

```
> head(mtcars)
```

```
          mpg cyl disp hp drat wt qsec vs am gear carb
Mazda RX4 21.0 6 160 110 3.90 2.620 16.46 0 1 4 4
Mazda RX4 Wag 21.0 6 160 110 3.90 2.875 17.02 0 1 4 4
Datsun 710 22.8 4 108 93 3.85 2.320 18.61 1 1 4 1
Hornet 4 Drive 21.4 6 258 110 3.08 3.215 19.44 1 0 3 1
Hornet Sportabout 18.7 8 360 175 3.15 3.440 17.02 0 0 3 2
Valiant 18.1 6 225 105 2.76 3.460 20.22 1 0 3 1
>
```

With the function str(), we can display the structure again:

```
> str(mtcars)
'data.frame': 32 obs. of 11 variables:
$ mpg : num 21 21 22.8 21.4 18.7 18.1 14.3 24.4 22.8 19.2 ...
$ cyl : num 6 6 4 6 8 6 8 4 4 6 ...
$ disp : num 160 160 108 258 360 ...
$ hp : num 110 110 93 110 175 105 245 62 95 123 ...
$ drat : num 3.9 3.9 3.85 3.08 3.15 2.76 3.21 3.69 3.92 3.92 ...
$ wt : num 2.62 2.88 2.32 3.21 3.44 ...
$ qsec : num 16.5 17 18.6 19.4 17 ...
$ vs : num 0 0 1 1 0 1 1 1 ...
$ am : num 1 1 0 0 0 0 0 0 ...
$ gear : num 4 4 3 3 3 4 4 ...
$ carb: num 4 4 1 1 2 1 4 2 2 4 ...
>
```

The function summary() outputs several statistics. I will go into the meaning of the individual statistics in the respective chapters (for space reasons,
the little brother of the data set mtcars is used here, cars):

```
> summary(cars)
     speed dist Min.
: 4.0 Min. : 2.00
1st Qu.:12.0 1st Qu.: 26
.00 Median :15.0 Median : 36.00
Mean :15.4 Mean : 42.98
3rd Qu.:19.0 3rd Qu.: 56.00
Max.   :25.0 Max.   :120.00
```

With this data frame we can now start the first analyses. To examine only
one column, enter the name of the data frame, a dollar sign, and then the
column to be examined.

```
> summary(cars$speed)
   Min. 1st Qu. Median Mean 3rd Qu. Max.
4.
0 12.0 15.0 15.4 19.0 25.0
>
```

Be careful here, because if you select only one column of a data frame, you will not necessarily get a data frame as a result. A frequent source of error! A *matrix* is like a data frame except that all cells in a matrix must have the same data type, which is a number. Some machine learning algorithms need a matrix as input.

A *factor* is a special vector where the characteristic values are predefined. For example, take a data frame where one column contains the gender of the respondents. The characteristics can be, for example, "male", "female" or "diverse". Now, these values can be stored in each cell as character. But since there is a limited number of different items, we could simply set these categories as such and thus save memory. Internally, vectors of type factor consist of integer values, each representing a category. These categories are called *levels* and their names are called *labels*. This term is also interpreted differently in machine learning, which I will then point out in each case. An example inspired by the R help; first, we create a vector:

```
> x <- c("Man", "Male", "Man", "Lady", "Female")
```

Then, a factor is created based on this vector:

```
> (xf <- factor(x, levels = c("Male", "Man" , "Lady", "Female"),
+ labels = c("Male", "Male", "Female", "Female")))
[1] Male Male Female Female
Levels: Male Female

> > str(xf)
 Factor w/ 2 levels "Male", "Female": 1 1 2 2
```

And here, the proof that internally integer values are used:

```
> typeof(xf)
[1] "integer"
>
```

A factor can also be expected as input by some machine learning algorithms.

4.3.4 Reading Data

R includes several exciting data sets, but most of the time, data must be read in externally. The most common case is reading in from a CSV file. CSV stands for Comma-Separated Values, although "comma" can be a bit misleading: It can just as well be a semicolon, or a tab. The good thing about CSV files is that they are usually pretty much like what we saw earlier in the definition of a data frame: each observation in a row, each column a variable. Thereby, the column name can already be included in the file.

Sometimes, it is useful to look at a file first to identify the separator. Smaller files can be easily opened and inspected using RStudio's file browser. With larger files from 5 megabytes, this is not possible. The built-in terminal in RStudio will help. Switch with the command

```
cd <FILE DIRECTORY>
```

to change directory to the directory that contains the file, and you can start using

```
head <FILE NAME>
```

to see the header of the file, similar to the head() command in R (see Figure 4.16).

RStudio makes importing files immensely easier, but again, understanding separators is important. To do this, you can click on a file in RStudio's file browser to decide whether to view the file or import it (see Figure 4.17). However, this only works if the selected file has the extension .csv or another extension that can be imported by RStudio. If you choose IMPORT data set, RStudio will use the Package readr that is part of the Tidyverse (see Chapter 5, "Exploratory Data Analysis"). If you want to import an Excel-file, RStudio will automatically select another appropriate package.

RStudio also offers two ways to read in CSV data via the ENVIRONMENT in the upper right corner:

- with base R

- with the library readr

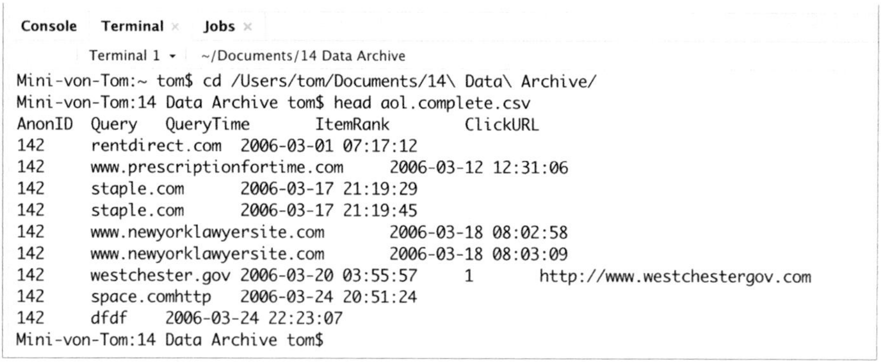

```
Console    Terminal ×   Jobs ×
         Terminal 1 ▾   ~/Documents/14 Data Archive
Mini-von-Tom:~ tom$ cd /Users/tom/Documents/14\ Data\ Archive/
Mini-von-Tom:14 Data Archive tom$ head aol.complete.csv
AnonID  Query   QueryTime       ItemRank        ClickURL
142     rentdirect.com  2006-03-01 07:17:12
142     www.prescriptionfortime.com     2006-03-12 12:31:06
142     staple.com      2006-03-17 21:19:29
142     staple.com      2006-03-17 21:19:45
142     www.newyorklawyersite.com       2006-03-18 08:02:58
142     www.newyorklawyersite.com       2006-03-18 08:03:09
142     westchester.gov 2006-03-20 03:55:57     1       http://www.westchestergov.com
142     space.comhttp   2006-03-24 20:51:24
142     dfdf    2006-03-24 22:23:07
Mini-von-Tom:14 Data Archive tom$
```

FIGURE 4.16

In the RStudio terminal, you can use regular UNIX commands to view the structure of a file with the "head" command and identify the separator. In this example it is a tab.

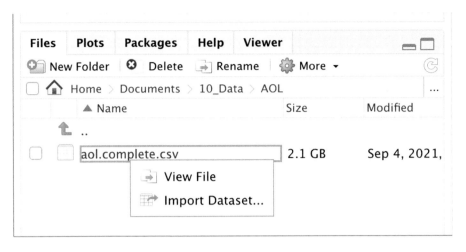

FIGURE 4.17
Importing data within the file browser of RStudio.

In addition, data imports from Excel, SPSS, SAS and Stata are offered (see Figure 4.18).

If you select the import via readr, a dialog box like the one shown in Figure 4.19. This dialog window offers several functions:

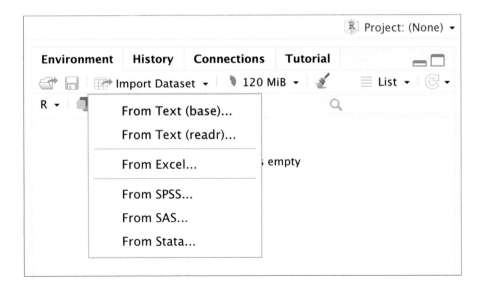

FIGURE 4.18
Reading data via the environment.

FIGURE 4.19
The dialog box for reading files into RStudio; in this example, the file seems to use a different separator than RStudio expects.

- By entering a number at SKIP, you can specify the number of lines to ignore at the beginning of the file.

- Some files do not have a "header" that describes the column. Then you have to disable FIRST ROW AS NAMES.

- Under DELIMITER, select the separator in the example from Figure 4.19 the separator must be changed; you can see the result in Figure 4.20.

- You can also specify which data type is involved for each column individually.

Let's take a closer look at this last item, because it regularly leads to challenges. In our example, the third column is a date with a time specification, but RStudio wants to import this column as a double. Ideally, we explain to RStudio right away how to read the format. To do this, click on the small triangle next to the guessed data type and then select the appropriate data format from the drop-down list (see Figure 4.21). In this example, it is Date-Time. Unfortunately, RStudio needs some help now – no wonder, because date formats cause a lot of confusion at any time.

We now need to specify how exactly the date format is constructed; in this case, first comes the year, then the month, then the day, all joined with hyphens, and then, separated only by a space, the time with a colon as a separator. In Figure 4.22 you can see how I specify the format here. Which is

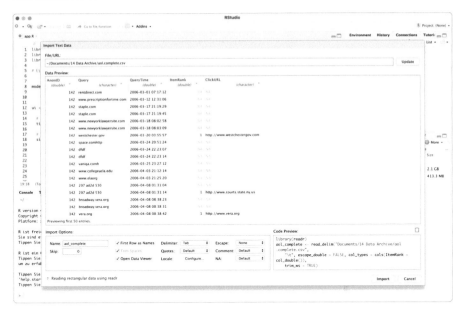

FIGURE 4.20
The dialog box for reading files into RStudio. Here I have changed the separator and the format now seems to be correct.

suboptimal, by the way: you don't see what the data actually looks like when you specify this format, so you have to remember what you're trying to do here. If you specify incorrectly, the column will remain empty or have only a few rows correctly recognized. In this case, however, everything should work.

The AOL data set

In case you were wondering what the data set is that you see in this section: It's the famous anonymized search queries that AOL published in 2006 and removed from the server shortly thereafter. That's because something went wrong with the anonymization process, so that some AOL users could be identified from their search queries. More than 36 million search queries are contained in the data set, and the unpacked data set is more than 2 gigabytes in size. You can find the data set on the web, unfortunately, for legal reasons, I cannot make it available for download on my website. A wonderful data set for data enthusiasts! For example, take a look at the searches of user 4417749. This user was identified and interviewed by journalists based on her searches, and she was still one of the users with harmless searches: https://alby.link/aolnyt.

As a R purist, you can of course just do it this way:

```
library(readr)
aol_complete <- read_delim("aol.complete.csv", "\t",
escape_double = FALSE, col_types = cols(QueryTime =
col_datetime(format = "%Y-%m-%d %H:%M:%S")), trim_ws = TRUE)
```

The second way to import data is to read in a URL to import. This actually works just like reading in a local file, except that you specify a URL instead of a file path. However, it is not recommended to load the files over the Internet every time. Instead, it's better to create a local copy after the first load, which you can then use for subsequent sessions. How to write data to your hard disk is described in Section 4.3.5.

As an alternative to the CSV-format, there are other formats, of which JSON has achieved a certain popularity. JSON stands for JavaScript Object Notation and can also handle data that is nested; the data does not have to be structured like a table. An example shows what a simple JSON file can look like:

```
{ "
firstName": "Jonathan", "
```

FIGURE 4.21
Configuration of the data type in RStudio data import.

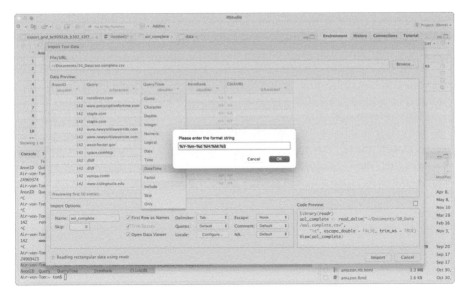

FIGURE 4.22
Format specification.

```
lastName": "Freeman", "
loginCount": 4, "
isWriter": true, "
worksWith":
[ "Spantree Technology Group", "InfoWorld" ], "
pets": [
{ "
name": "Tom", "
type": "Mouse"
},
{ "
name": "Jerry", "
type": "Cat"
}

    ]
}
```

In CSV format, this would be a bit more complicated to display, because there are two entries for the pets. Since the number of pets can vary for each entry, it makes little sense to provide columns in advance just for the pets (pet1_name, pet1_type, pet2_name, pet2_type). There are always outliers that have more pets than usual. One solution could be to simply name all pets in one column, separated by commas, and still interpret that column separately, but that adds complexity. JSON can have its advantages.

The use of JSON is questionable for many data sets that could simply be stored in CSV format. On the one hand, JSON requires additional memory, both on the hard disk and in RAM, and is then also dependent on an interpreter. For another, a JSON file usually needs to be loaded completely so that it can be interpreted. R cannot read JSON files without a package; for this, the Tidyverse currently uses the package jsonlite in addition to the package readr. Unfortunately, there is not only a JSON format, but also the NDJSON format (Newline-Delimited JSON), which leads to error messages even if the code is correct. Since this is a special case, I refer to my blog for explorations with JSON in R.

For special applications like Spark (see Chapter 10, "Workflows and Tools"), other file formats may be more advantageous. Otherwise, readr installs other packages, including:

- DBI for databases

- haven for SAS, SPSS and Stata

- httr for APIs that can be reached via the web

- readxl for reading and writing Excel tables

- rvest for scraping of data from the web (so HTML code is also interpreted here)

- xml2 for XML-files

Create PowerPoint presentations with R

There are also packages for writing Word and PowerPoint files. In fact, I've written PowerPoint presentations completely automated in R before to reduce the monthly work for a report that the client really wanted in PowerPoint format. It takes a bit of time to get the formatting aesthetically accident-free, though. You can learn more about this on the website for the book.

No matter what format, data is often not clean. Even if the import, for example via the RStudio import menu, seems to work at first, there may still be problems. This is what such a problem may look like:

```
> library(readr)
> aol_complete <- read_delim("aol.complete.csv",
+ "
```

```
\t", escape_double = FALSE, trim_ws = TRUE
)
-- Column specification ------------------------------------
----------------------
cols(
AnonID = col_double(),
Query = col_character(),
QueryTime = col_datetime(format = ""),
  ItemRank = col_double(),
ClickURL = col_character()
)
|=========================================================
========| 100% 2173 MB
Warning: 29 parsing failures.
row col expected actual file
3558412 AnonID a double AnonID 'aol.complete.csv'
3558412 QueryTime date like QueryTime 'aol.
complete.csv'
3558412 ItemRank a double ItemRank 'aol.complete.csv'
7172919 AnonID a double AnonID 'aol.complete.csv'
7172919 QueryTime date like QueryTime 'aol.complete.csv'
..........................................................
...............
  See problems(...) for more details.
```

Sometimes, a row contains characters that cause an entry in one column to be split in two columns, resulting in more columns than were actually expected. In this case, data also seems to be mixed up. It is offered to get all details with problems(...). In the parenthesis, we enter the object name used to write the result of the import to:

```
> problems(aol_complete)
# A tibble: 29 × 5
row col expected actual file
<int> <chr> <chr> <chr>
1 3558412 AnonID "a double" AnonID 'aol.complete.csv'
2 3558412 QueryTime "date like " QueryTime 'aol.complete.csv'
3 3558412 ItemRank "
a double" ItemRank 'aol.complete.csv'
4 7172919 AnonID "
a double" AnonID 'aol.complete.
csv'
5 7172919 QueryTime "date like " QueryTime 'aol.complete.csv'
6 7172919 ItemRank "a double" ItemRank 'aol.complete.csv'
7 10853833 AnonID "
```

	ISBN	Book–Title	Book–Author	Year–Of–Publication	Publisher	Image
6447	0751516678	Hester Dark	Emma Blair	2002	Warner Futura	http://
6448	0297828991	Mr Foreigner	Matthew Kneale	2003	Trafalgar Square	http://
6449	0061013722	The Clairvoyant	Marian Thurm	1999	HarperTorch	http://
6450	0451163540	Triple	Ken Follett	1991	Signet Book	http://
6451	0735201994	"Peterman Rides Again: Adventures Continue with the…	the Catalog Business"		2000	Prentic
6452	1414011458	No More Bobs	"Peterman Rides Again:	2003	Authorhouse	http://
6453	1564029654	Owl Babies	Adventures Continue with the Real \"J. addell	1996	Candlewick Press (MA)	http://
6454	0307121259	I Am a Bunny (Golden Sturdy Book)	Peterman\" Through Life &	1963	Golden Books	http://
6455	0671493183	Doggies (Boynton Board Books (Simon & Schuster))	Sandra Boynton	1984	Little Simon	http://
6456	0679882812	Dr. Seuss's ABC: An Amazing Alphabet Book! (Bright a…	DR SEUSS	1996	Random House Books for Young Readers	http://
6457	067144901X	Moo Baa La La La	Sandra Boynton	1982	Little Simon	http://
6458	0399215921	The Very Busy Spider	Eric Carle	1989	Philomel Books	http://
6459	0394900189	Are You My Mother (Beginner Books)	P. D. Eastman	1966	Random House Books for Young Readers	http://
6460	0394800206	Go, Dog, Go (I Can Read It All by Myself Beginner Boo…	Philip D. Eastman	1961	Random House Children's Books	http://
6461	0205150227	Curious George (Sandpiper Books)	H. A. Rey	1973	Houghton Mifflin	http://

FIGURE 4.23
Problematic row after an import, here with the cursor on the cell and a pop-up.

```
a double" AnonID 'aol.complete.
csv'
8 10853833 QueryTime "date like " QueryTime 'aol.complete.csv'
9 10853833 ItemRank "a double
" ItemRank 'aol.complete.csv'
10 14504524 AnonID "a double
" AnonID 'aol.complete.csv'
# ... with 19 more rows
>
```

The problem is not completely visible here; also, not all lines are displayed, so it is recommended to have a closer look at the lines. You can do that in this case with the command print(problems(aol_complete), n=29). Another possibility is to look at the rows in RStudio in the viewer. If you move the cursor to a cell in the data set, a pop-up will appear showing the whole row, because in the viewer itself the displayed length is limited (see Figure 4.23). This is not always easy, however, because if you have millions of rows, it may take time to scroll to the appropriate row. You can also use R itself to get to the line, for example like this:

```
> aol_complete[3558412,]
# A tibble: 1 × 5
AnonID Query QueryTime ItemRank ClickURL
<dbl> <chr> <dttm> <dbl> <chr>
1 NA Query NA NA ClickURL
```

Alternatively, you can simply examine the line in the terminal in a good text editor that can handle large files. This is often more advisable, since you can't make sure that RStudio imported everything in the first place. Also, it's quick: Instead of looking for line 7172919, you need to add the number of header lines, in this case one, so we end up with line 7172920:

```
Air-von-Tom:~ tom$ awk 'FNR==7172920' aol.complete.csv
AnonID Query QueryTime ItemRank ClickURL
```

A simple case: Here, apparently the first line has been repeated, because the original files came from several CSV files, and these were simply strung together.

What exactly is a text editor?

Microsoft Word and LibreOffice are not text editors, but word processors. These don't show you the raw data – it is already a format. Text editors, on the other hand, display the raw data. If you don't want to use the classics vi, nano, or Emacs, then Microsofts Visual Studio Code is a good option.

Quite often, however, such problems are more difficult to solve. For example, if every value of a column is set in quotes, but an entry also has quotes in a column, which are *escaped* with a backslash. However, RStudio does not know this at first and thus imports the rows incorrectly. Fortunately, this can also be fixed by specifying the backslash as an option in ESCAPE IN the import menu.

Unfortunately, this is not always so easy. Sometimes, data is so unclean that all the tricks are of no use, and individual lines remain unclean. Of course, the original file can then be cleaned manually. For a one-time import, this may be an option. But as soon as you want to build a model that imports new data frequently or even continuously, this way is not viable. Then, either YOU have to write other routines that check the data for errors and compensate for them or - and this may also be an option worth considering - you can accept the errors because they do not affect the model and only a small proportion of the data is affected. But this decision should be made and documented based on data. It should also be noted that the number of affected data MAY BE small, but this may be different in the next data delivery.

A clean data import is the foundation for successful analysis and application of machine learning models. Even though tools like RStudio make importing data easier, that doesn't mean your brain can be turned off. In Chapter 5, "Exploratory Data Analysis," I will cover the topic of data quality in more detail.

4.3.5 Writing Data

You can write a simple CSV file with Base R like this:

```
write.csv(books,file="books.csv")
```

A common mistake is not taking into account which working directory you are currently in, and then not finding the file you just created. With the command getwd() (get working directory), you can see where you are. With setwd() (set working directory), you define another working directory. If you are working in a project in RStudio, a working directory other than the default RStudio directory is already selected.

4.3.6   Shortcuts

Code should always be commented, even if you like to believe that the code is self-explanatory when programming. When you open your code again after a few months or years, the lack of comments takes its revenge. Comments are marked with a hash; code in them is not executed, as in this example:

```
> print("Hello world") # This is a great line of code
```

You should choose object names in such a way that people can understand what they are about. A variable with the name "a" is mostly of little use, and after a few weeks one has forgotten what it actually stood for. There are a few style guides, within an organization, a common style guide should be followed. Hadley Wickham, the creator of the Tidyverse, has published a wonderful style guide at https://style.Tidyverse.org. The main points from this:

- Variable and function names should be lowercase throughout and parts should be connected with an underscore if necessary (the so-called snake case): result_transformation is an example.

- Variable names should be nouns; function names should be verbs.

Snake versus camel case

Besides the snake case (great_variable) there is also the camel case (greatVariable) and the possibility to separate everything with dots (great.variable). The last variant is used by R for function and class names, so you should stay away from it when programming.

As described earlier, R supplements commands. This goes so far as to complement more than just parentheses and quotes. For example, if you enter only "sum" and then press the Tab key, both R and RStudio offer suggestions.

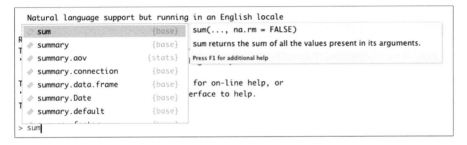

FIGURE 4.24
A suggestion generated by R, which command can be used here with which
parameters.

As a special service, R then also shows with which parameters the respective
command can be issued, as can be seen in Figure 4.24.

Often, you use a command more than once during an analysis. If you use
the up and down arrow keys, you will see the commands you have already
entered in the console. With a little practice, you can speed up your work in
R considerably with these tips.

If a command is not complete, R kindly does not throw an error message,
but indicates by the + sign that you should please add something:

```
> 2+
+
```

You can enter the still missing part of a command. If you press Enter,
R still waits for an input. The only way to get out of this is to use the key
combination Ctrl+C. If you need help, a ? in front of the respective command
will help:

```
?summary
```

This will either open a new window or, in RStudio, open the Help tab
and the appropriate help page. I strongly recommend the use of spaces in the
code, for example:

```
> a + b
```

because especially with longer lines of code, this is more readable than

```
> a+b
```

As already mentioned, however, you may not use spaces in variables; even
within the assignment character $< -$, a space is not allowed. Exceptions con-
firm the rule here, too, because as soon as you put a variable in quotation
marks, this also works: 'great variable.' However, this is not recommended!

5

Exploratory Data Analysis

Did you hear the one about the statistician who drowned trying to wade across the river? He knew it was three feet deep... on average (Old statistician joke, author unknown)

An exploratory data analysis is an approach from statistics described by Fahrmeir et al. as follows [23], own translation:

Beyond the presentation of data, [exploratory data analysis] is designed to search for structures and features in the data and thus can often lead to new questions or hypotheses in the respective applications. It is therefore typically used when the research question is not well defined or even the choice of an appropriate statistical model is unclear.

Before we venture into machine learning models, the available data must first be cleaned, examined, and, if necessary, transformed. This is usually done in the context of an exploratory data analysis, or EDA for short. Even though Fahrmeir et al. say that the research question does not have to be precisely defined, it is nevertheless advisable to define at least one initial research question as clearly as possible at the beginning of an analysis and to return to this initial research question again and again before "getting lost" in an analysis.

At the same time, an EDA is a kind of "state of mind," as Hadley Wickham and Garrett Grolemund describe it: Every idea that comes to you should be pursued in the analysis process. It's a creative process, so it's also okay to generate a lot of questions because the most difficult thing is often to ask the right questions of the data:

Data exploration is the art of looking at your data, rapidly generating hypotheses, quickly testing them, then repeating again and again and again. The goal of data exploration is to generate many promising leads that you can later explore in more depth. [30]

Exploratory data analysis was "invented" by John W. Tukey in his article "The Future of Data Analysis" [47]; his book *Exploratory Data Analysis* is considered a classic and the basis for the topic of Data Science [48].

DOI: 10.1201/9781003426363-5

Exploratory Data Analysis or Feature Engineering?

While the concept of exploratory data analysis comes from statistics, the term *feature engineering* comes from machine learning. The boundaries are blurred because in both approaches, data is explored to understand, clean, and transform the data to identify the variables that can later be used in a model. However, in exploratory data analysis, I see the focus on the first steps, and in feature engineering, on the last.

We will approach EDAs in several steps:

- What should be considered when collecting, cleaning, and transforming data?

- Then, we will look at notebooks that we will use for exploratory data analysis.

- You will learn about an R package that simplifies exploratory data analysis immensely, the Tidyverse.

- Then, we visualize data.

- Finally, we analyze data once as an example, introducing some important concepts of statistics.

This chapter could be a book of its own; where I need to be more brief, I will refer to the relevant further reading.

5.1 Data: Collection, Cleaning and Transformation

Once the requirements and questions for a Data Science project have been determined, the first major effort comes, which usually takes a lot of time. Data is not always available in the way you need it, and even more rarely is it "clean." Let's think back to the AOL data set from the previous chapter for a moment: if we want to know whether the length of a search query has an impact on search success, this information is hidden in the search query column. We have to extract this information from the variable first. At the same time, we then notice that some users have extremely long search queries with dozens of words. Should these queries really be included in such a consideration?

The preparation of data for use in a model is often underestimated. It is not uncommon for the effort to be 80% of the total time spent on a project.

5.1.1 Data Acquisition

Moreover, data is not always be available even though it is in theory, for example for data protection reasons (see Chapter 11, "Ethical Handling of Data and Algorithms") or because data was previously collected without specific consent. Furthermore, a distinction is made between qualitative and quantitative data. Data from open interviews are qualitative in nature; data from measurements or surveys are quantitative. If, for example, you were asked which party you would vote for if federal elections were held next Sunday, the possible answers would be countable and thus quantitative.

However, if the question were to ask why you would vote for a particular party without giving you any answer options in advance, then the evaluator could paraphrase and categorize the answers and thus make them countable, but programs could not easily do so. Obviously, it is difficult to analyze such data from interviews unless it involves text analysis using computational linguistic methods. It is different with quantitative data, which are usually available in such a way that they can be analyzed by machines without a text analysis. Data are therefore not all the same, which is not always easy to understand, especially for outsiders.

By selecting a data source, a (hopefully unintentional) bias may already be present. For example, if I conduct a survey on political party preference in Manhattan, the answers will probably be different from those in Staten Island. Sometimes, data is collected in such a way that you don't even realize that the data is biased. Example: You take a look at how many users come to your website and discover that virtually no one uses a cell phone for this purpose. You conclude that your topic is not attractive for mobile users and stop optimizing your site for mobile. But, it could just as easily be that your site just doesn't read well on mobile devices, and that's why Google doesn't show it on the first search results page. This survivorship bias, as it's called, results in looking only at the cases that somehow managed to make it into the investigation, so the wrong conclusions are drawn.

Data acquisition is a separate area that can vary greatly depending on the industry and topic. I refer to the specific technical literature in each case.

5.1.2 How Much Data Is Enough?

This is one of the most common questions at the beginning of a Data Science project. It is not easy to answer, because depending on the algorithm, the answer may be different. In general, the more data, the better. Some algorithms, like Deep Learning, tend to require large amounts of data, whereas Naive Bayes can, but need not, produce robust results with just a small amount of data. If you think back to Chapter 3, "The Anatomy of a Data Science Project," and the process of testing, training, and validating data presented in it, it also becomes clear that a few dozen data sets is not enough. Popular tutorials such as predicting which Titanic passenger will survive are based on just over 1,000 passengers, and clustering with the 150 rows of data from the Iris data set is also often found on the Internet. But, this is the exception. A few thousand observations are always a good start, but this may also mean that data also needs to be labelled manually.

In this context, the problem of class imbalance should be mentioned again. As explained before, a class imbalance occurs when the individual classes have a greatly different size. Oversampling can help, i.e. drawing a sample from the smaller class several times to generate more data. Likewise, the larger class could be undersampled, i.e. take a sample from the large class instead of the whole data set. In a sample, it is important that the elements are chosen randomly; for example, not the first 20 rows of a data set, but randomly selected rows.

There are also more complex methods such as synthetic samples, for which new data is generated from the attributes of the smaller class. However, all of this must be taken with a grain of salt. An example here is so-called fraud, i.e. fraudulent business transactions, whether because one is sold something that is either delivered incorrectly or not at all, or because one is sold something that is delivered but never paid for. Fortunately, only a small percentage of business transactions are fraudulent. And that's what can be a problem for machine learning. After all, in supervised learning, if a class is very rare, then there is little chance that the machine will be good at detecting those cases. All of this is already apparent in the exploratory data analysis and can be remedied, if necessary, by the methods described.

5.1.3 Data Cleaning: The Different Dimensions of Data Quality

Depending on the author, a different number of dimensions of data quality are distinguished (see, for example, Loshin, 2019 [35]), some of which will be described here:

- Validity: Do the values comply with the required standards? For example, if a date and time value is required, but the time value contains only hours from 0 to 11, then the standard of an international time specification is not met.

- Accuracy/Accuracy: Are the contents of the data correct? Example: In a data set, a person is over 200 years old. This is not only improbable, but at least for the moment, still impossible.

- Consistency: Is the data consistent across multiple systems? Example: There is a customer database and another database with the payment experiences with these customers. If data such as name and address are not consistent, it will not be possible to synchronize the data.

- Relevance: Does the data reflect the interests of the users? While there is a lot of data, it does not always contain what we are really interested in.

- Integrity: Structural or relational quality, e.g. de-duplication so that no data points are duplicated.

- Accessibility: How easy is it to get to the data when you need it?

- Timeliness/Currency: To what extent does the data reflect the current data situation? Example: Credit default data from before the 2008/2009 financial crisis can only be used at later points in time with a great deal of caution, simply because the world has changed a lot since then. However, the gap between the creation and availability of data can also be critical, in that real-time applications rely on data being available in the system as quickly as possible.

- Uniqueness: Each entity exists only once in the data, at least with the respective attributes.

- Completeness: All data, or at least a representative sample, is available for the required investigation.

However, it is not always that simple, especially when it comes to correctness. The case of Hadlum vs. Hadlum is often mentioned in the literature, see especially [9]. Mrs. Hadlum gave birth to a child 349 days after her husband left for war. The typical pregnancy lasts 266 days. With such outliers, the following questions should be asked:

- Is this value typical?

- Is it unusual?

- Is it impossible?

A pregnancy can go longer than 38 weeks, but more than 10 weeks over? Mrs. Hadlum won in court, which said that such a long pregnancy, while very unlikely, was still possible. Obviously, expertise is needed to make a decision in such cases as to how accurate the data actually is.

5.1.4 Data Transformation: The Underestimated Effort

The ideal data set for most analytics and machine learning algorithms, as mentioned previously, has each observation in a row and each variable in a column. This is not always the case. And sometimes, the data is in multiple files, which then need to be merged into one file.

Some machine learning algorithms require numbers as input because distances are often used in these algorithms. Examples are k-Means and Support Vector Machines. However, not all data is always in the form of numbers. For example, a column in a data set might contain the gender of individuals, and this might be diverse, male, or female. This is a so-called categorical variable. At first glance, one might be tempted to assign a number to each gender, for example, diverse a 1, male a 2, and female a 3. But at a second glance, this makes no sense because when it comes to distances, the question is whether the female gender is actually further away from the gender "diverse" than the male gender. Distances do not make sense here, and thus, it is not purposeful to assign a number to each gender.

Data Scientists have found a different solution for this, which I will explain in detail in Section 7.1.4. Each gender gets its own column, and if the gender applies, a 1 is found in the respective cell, otherwise not. Strictly speaking, however, not every gender gets its own column, because the third column results from the two columns before. If someone is not diverse or male, then this person can only be female.

Such a transformation is only one example of many other necessary transformations. Not all algorithms require numbers; Naive Bayes, for example, prefers categories. How do you make categories out of numbers, for example, when it comes to income? One could make categories of income, but how to make the bounds for each category? Not an easy decision, and first an analysis is needed to form meaningful groups.

By the way, you should beware of cleaning data in Excel first, unless your data is much too large for that because once data has been opened in Excel and saved again, the data may be different, even if a CSV file is exported from Excel. And, this does not only concern the sometimes unwanted conversion to a date format.

5.2 Notebooks

Now that you have learned something about data cleaning and transformation, at least theoretically, I would like to introduce you to another tool that simplifies the work of exploratory data analysis immensely: Notebooks.

5.2.1 EDAs with Notebooks and Markdown

Notebooks are a great invention for data analysis. They exist for other languages as well. The JuPyteR project, for example, was developed for Julia, Python and R, hence the name. In a notebook, text, the actual software code, and data and results are brought together in one document. For example, the analyst first writes down what the goal of the analysis is; then, the appropriate code is written, and the result is calculated in the document, and eventually interpreted in a text (see the standard notebook in RStudio in Figure 5.1).

By having everything together in one document, the approach and software can be replicated by other analysts, and analyses can be shared more easily. Reproducibility of results is a big issue in the scientific world, but also increasingly in the Data Science scene. Unless precluded by data protection

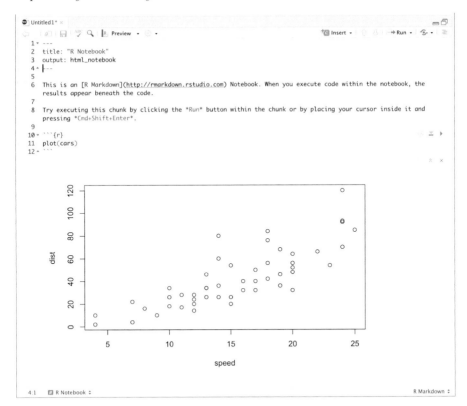

FIGURE 5.1
Example of the default notebook in RStudio.

laws or other restrictions, data, software, and documentation can be easily shared in this way.

To further illustrate the genius of notebooks, imagine that you have examined a large data set in Excel. There are many formulas in your spreadsheets, and perhaps some text. But to some extent, you have had to modify the data, for example a date value, to make the analysis work. After three months, would you still know exactly what you were doing? And if you sent the analysis to someone on your team, would they be able to follow your thoughts? Very likely not.

With a notebook, it's different. You start with an empty notebook, then import the data (documenting at the same time how you did it), clean and transform it, analyze it, and finally apply models. All in one document, the whole workflow for you and your colleagues to review. Take a look at an example at https://tom.alby.de/analysis/nycflights.html.

A notebook is created in RStudio by opening the menu as explained in Chapter 4, "Introduction to R", and selecting R NOTEBOOK there. When

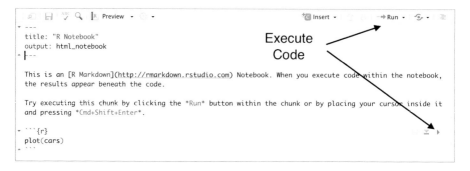

FIGURE 5.2
Running code in a code block.

a new notebook is created, a piece of code is always already present in a code block called a *chunk*:

```
plot(cars)
```

If you execute the command, the plot will appear directly below the code block, which can then be described further in the body text. The code is executed either by clicking on the small triangle in the upper right corner of the code block or by pressing Ctrl/Command+Enter, depending on the operating system (see Figure 5.2). In this way, one part of the analysis follows the other, and a comprehensible walk through the thoughts of the author of the notebook results.

You insert a new code block either by clicking INSERT in the interface and then on the respective language, or by pressing a key combination that depends on the respective operating system (see Figure 5.3). RStudio allows you not only to insert R code, but also to use other languages such as Python. So, in one document, pieces of code of different languages can be used, and they can exchange data with each other. More than one line of code can be written in a code block, but only in the same language. The key combination for executing the code then only applies to the respective line, not to the entire code block.

This is also a common source of errors: A code block starts with three quotes and an r in curly brackets and ends again with three quotes:

```
'''{r}
plot(cars)
'''
```

Due to incorrect copying or carelessness, it can happen that part of the beginning or end of a block is missing. The green arrow in the upper right corner of the code block disappears, so that the code is no longer executable. Nothing happens at all.

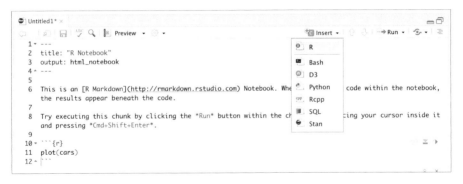

FIGURE 5.3
Insert a new code block.

Notebooks are based on Markdown, a lightweight markup language for formatting text. Instead of marking a text section as in Word and then setting it in italics via a menu command, simple commands are placed in front of or around a text section in Markdown. A whole book could be written about Markdown, but only the most important commands will be explained here.

Markup versus Markdown

You will certainly have heard of markup before, for example with HTML, the Hyper-Text Markup Language. So what exactly is Markdown? Markdown is actually just another markup language.

As mentioned in Chapter 4, "Introduction to R," the hash character is actually intended for comments, but in notebooks it takes on a different meaning:

```
# I am a big headline ##
I am a slightly smaller headline
### And I am an even smaller headline
An unnumbered enumeration is generated in this way:
* The first point
* The second point
```

The so-called Front Matter of a Markdown document describes how to deal with individual elements of the document:

```
---
title: "Example notebook"
output:
```

```
html_document:
toc: true
df_print: paged
---
```

This example defines what the title of the document is, that a table of contents should be generated from the headings, and that the data frames within a document should not be output completely on one page but with pagination. Important: If you make a mistake in the Front Matter, the whole notebook may not work afterwards or cannot be converted into another document (see the next section).

The result after exporting to HTML (called *knitting*, see the next section) then looks like the notebook in Figure 5.4.

By the way, every time you save your notebook, an HTML file is created or an existing HTML file updated. This can be a bit annoying, because actually, you don't always need these HTML files. You can prevent this behavior by deleting this expression in the Front Matter:

FIGURE 5.4
The output of the notebook after knitting.

```
output: html_notebook
```

RStudio, in the version that is available at the time of writing (1.3.1093), has a somewhat confusing way of dealing with notebooks. This is because when you create a new document, you will find both the R NOTEBOOK and R MARKDOWN DOCUMENT ITEMS IN the menu. In both cases, a document with the extension .Rmd is created. The only difference is that when you select R MARKDOWN DOCUMENT, a dialog box helps you define certain properties of the target document, and in the process, some of the front matter is already specified. At https://alby.link/notebook you will find a video with a tour of a notebook in RStudio.

5.2.2 Knitting

Notebooks can be exported to other formats, as shown in the previous section, for example to HTML pages or to a PDF document. With Markdown, specific commands can be used to change formatting, insert HTML links, and so on. Here the data, if included in the notebook, can also be seen in the resulting document. This can cause these documents to become very large, so some brain power needs to be invested up front as to what data is really essential to share. An example of exploratory data analysis created with a notebook and Markdown can be found at https://alby.link/eda.

But, not only can single HTML pages be created, but also whole websites. For example, my website is written entirely in R, and each page is a Markdown document. (See Figure 5.5) I then use the Hugo framework to convert these Markdown documents into static HTML pages.

Quarto

Quarto is the next-generation way of creating reproducable documents. While Rmd notebooks require R, Quarto notebooks work with different languages. You will find more information at http://quarto.org

---

5.3 The Tidyverse

Now that you've just learned about a tool in which you can document your exploratory data analysis, I'll introduce you to a dialect of R that will greatly simplify your work.

FIGURE 5.5
Creating your own website with Markdown and Hugo in R.

5.3.1   Why Use the Tidyverse?

The Tidyverse is a package for R penned by Hadley Wickham, the Chief Data Scientist of RStudio, who has also written many other packages for R. Jokingly, the Tidyverse is sometimes referred to as the *Hadleyverse* by Hadley's followers. In fact, the Tidyverse is not just a single package, but a collection of packages.

There are other extensions for data transformation, especially data.table should be mentioned here. In the classroom, however, Tidyverse has proven to be easier, because in principle, you can express yourself with Tidyverse in the way you think.[1] This will be discussed in more detail in the next sections.

The Tidyverse contains several libraries:

- ggplot2: extended plotting possibilities (see Section 5.4.4)

- dplyr: a popular library for data manipulation

- tidyr: to "clean up" data

- readr: to read in data

- purrr: extension of functional programming in R

---

[1] This is also the title of a talk that Wickham gave, you can still find it on YouTube.

```
> library(tidyverse)
Registered S3 methods overwritten by 'dbplyr':
  method          from
  print.tbl_lazy
  print.tbl_sql
— Attaching packages ——————————————————————— tidyverse 1.3.1 —
✓ ggplot2 3.3.3    ✓ purrr   0.3.4
✓ tibble  3.1.2    ✓ dplyr   1.0.6
✓ tidyr   1.1.3    ✓ stringr 1.4.0
✓ readr   1.4.0    ✓ forcats 0.5.1
— Conflicts ————————————————————————— tidyverse_conflicts() —
x dplyr::filter() masks stats::filter()
x dplyr::lag()    masks stats::lag()
>
```

FIGURE 5.6
Confusing messages after loading the Tidyverse library.

- tibble: a modern version of data frames; when the Tidyverse is loaded, data frames automatically become tibbles.

- stringr: library for editing strings

- forcats: library for solving problems with Factors

It is possible that also only single libraries are installed; with

```
install.packages(tidyverse)
```

all libraries are installed at once and started with

```
library(tidyverse)
```

Unfortunately, the Tidyverse displays several messages directly after loading, confusing especially beginners (see Figure 5.6).

Some lines are in red, which causes many participants in my courses to stop working because they assume that this is an error message. Then, there is something about "Conflicts", which adds even more to the confusion.

In fact, nothing bad has happened here. Strangely enough, the Tidyverse shows in the signal color red which methods are "overwritten". In the case of conflicts, it simply says that if you now enter filter(), for example, this function will now behave differently than before.

A common source of errors arises from the fact that some packages offer a function that already exists in Base R or in another package. Thus, if R is started first, then, for example, the function filter() works as it was intended by the creators of R. If then, for example, the Tidyverse is loaded together with its package dplyr, then filter() suddenly behaves quite differently, namely, as it is intended in dplyr. Although the Tidyverse warns about this after loading, in the course of a project, you can forget something like this. For example, if you normally want to use the filter() function as is and then use the Tidyverse variant for only one command, it is not worth loading the Tidyverse package first, using the other variant of filter(), and then unloading the package again. One way to use the Tidyverse variant without loading the package is to enter the command as follows:

```
> dplyr::filter(mtcars, cyl==6)
          mpg cyl disp hp drat wt qsec vs am gear carb
Mazda RX4 21.0 6 160.0 110 3.90 2.620 16.46 0 1 4 4
Mazda RX4 Wag 21.0 6 160.0 110 3.90 2.875 17.02 0 1 4 4
Hornet 4 Drive 21.4 6 258.0 110 3.08 3.215 19.44 1 0 3 1
Valiant 18.1 6 225.0 105 2.76 3.460 20.22 1 0 3 1
Merc 280 19.2 6 167.6 123 3.92 3.440 18.30 1 0 4 4
Merc 280C 17.8 6 167.6 123 3.92 3.440 18.90 1 0 4 4
Ferrari Dino 19.7 6 145.0 175 3.62 2.770 15.50 0 1 5 6
>
```

Here you call the command from dplyr but do not load the library. This is a very useful method, as you will surely find out in future projects.

5.3.2   The Basic Tidyverse Verbs

dplyr uses five main verbs that allow you to execute basic commands:

- select(): to select variables or columns, respectively

- filter(): to select rows

- mutate(): to create new variables/columns or also to modify existing columns.

- summarise(): calculate results

- arrange(): to sort rows

An elementary approach in the Tidyverse is the so-called *pipe*. Under UNIX, pipes are a popular way to use the output from one function as input for another function, e.g.

```
grep "2018" report.csv | wc -l
```

In this example under UNIX, the file report.csv is searched for all lines containing the character string 2018. The output would then be exactly these lines. With the operator | this output is directed to the next command. wc -l counts the lines (wc stands for word count; with the parameter l, lines and not words are counted). We find out how many lines contain the string 2018 with this command. Pipes are a very efficient means under UNIX.

You can think of it like a cooking recipe:

```
Eggs %>%
    Break(Eggs) %>%
    PourInPan() %>%
    ScrambleUm() %>%
    ServeScrambled()
```

The result of the first step is passed to the next step. This is exactly the mechanism the Tidyverse uses, focusing on readability, not efficiency as in UNIX systems.

In dplyr, the pipe symbol

```
%>\%
```

is used instead of the Unix pipe symobol. Assuming that report.csv is already loaded and the corresponding data frame is named "report" and looks like this:

```
> report
# A tibble: 4 x 2
year name
<dbl> <chr>
1 2017 Anna
2 2018 Berta
3 2018 Carla
4 2016 Dieter
>
```

Then the command to find and count all entries with the year 2018 in the Tidyverse would look like this:

```
report %>%
    filter(year == 2018) %>%
    summarise(number = n())
```

These lines of code deserve a detailed explanation, as they are a wonderful introduction to the concept of dplyr. Each line contains exactly one instruction, so as in a cooking recipe, each line guides you through the sequence. One could express these three lines in natural language as follows:

1. Show me the record report.

2. From this, show me only the rows that have the value "2018" in the "Year" column.

3. Count the resulting rows, and show them to me in a new variable number.

The first two lines are probably self-explanatory, but line 3 needs some explanation. The summarise() function takes parameters that define how to summarize. n() is a function that simply counts the number of lines. By the way, you can write both summarise() and summarize().

---

n = n()? And when do I use = and when ==?

In many tutorials on the net, summarise(n = n()) is used as an example, where the first n is the variable and the second n() is the function, so the two n have nothing to do with each other. From experience, this does not exactly simplify understanding in courses, so I have chosen a different example here, but I still want to mention the common notation as well, since you will come across it sooner or later.

In Chapter 4, "Introduction to R", the difference between $<-$ and $=$ was discussed. In the previous example, you can now see two different ways to use equals signs. In the first example, the contents of the variable year are compared to the value 2018. A double equals sign is always used in a comparison. In the second example, number $=$ n(), no $<-$ is used, even though a new object is created. This is because this happens inside a function (summarise), and then, the equal sign is used.

Other functions are, for example, mean() for the mean, sd() for the standard deviation, or min() and max() for the range.

From this example it becomes clear what the strength of dplyr is: programmers have to think about the steps that are necessary to get to a certain result. This process, which normally takes place in the mind, can now be broken down very easily in code. This is what Hadley meant when he said that you can express yourself in R.

It is not necessary to indent from the second line as in the example. Although this happens automatically when you press Enter after writing a line, it could actually be written this way:

```
report %>% filter(year == 2018) %>% summarise(number = n())
```

Obviously, the charm of coding in dplyr, namely, the readability and thus improved traceability of the code, is partially sacrificed when the formatting is removed.

Of course, it would have been possible to arrive at the result without dplyr. For example, the R code could have looked like this:

```
nrow(report[report$year==2018,])
```

The expression can be expressed in natural language as follows: Count the number of rows in the data frame "report" where the column "year" in the data frame "report" has the value 2018. The individual steps in this example, which are necessary to arrive at the same result, are nested and not easy to understand, especially for R beginners. dplyr simplifies programming enormously.

One more thought about the filter() command: Again, if we think about it more carefully, we can see an imprecision. When you filter something, do you filter "out" what is specified so that it is no longer in, or do you filter it "in," so to speak, so that only what is specified remains in, as in the example? In fact, filter() doesn't filter out; it filters in. So when filter(year==2018) is executed, it only searches for the entries with "2018," not all entries without "2018." This can also be misleading at the beginning.

5.3.3 From Data Frames to Tibbles

A special feature of the Tidyverse is the tibble. If you have been paying close attention, you will have seen this term before, when we reviewed the data frame report. It was no longer a data frame, but a tibble, what Hadley Wickham calls a modern form of data frames. See https://tibble.tidyverse.org for more information. For our work here, however, tibbles have no significant impact for now. Once you have loaded the Tidyverse, each Data Frame automatically becomes a tibble.

5.3.4 Data Transformation

As an example, we use the data set nycflights13 from the package of the same name. It contains all outbound flights from New York airports from 2013, as well as additional data on the airlines, the aircraft, and the weather that prevailed at the airport in question at the time of the flights. This data set must first be installed and loaded:

```
install.packages("nycflights13")
library(nycflights13)
```

It consists of several tables or data frames or tibbles:

- flights

- planes

- airlines

- weather

- airports

Let's look at flights:

```
> flights
# A tibble: 336,776 x 19
year month day dep_time sched_dep_time dep_delay arr_time
<int> <int> <int> <int> <dbl> <int>
1 2013 1 1 517 515 2 830
2 2013 1 1 533 529 4 850
3 2013 1 1 542 540 2 923
4 2013 1 1 544 545 -1 1004
5 2013 1 1 554 600 -6 812
6 2013 1 1 554 558 -4 740
7 2013 1 1 555 600 -5 913
8 2013 1 1 557 600 -3 709 9 2013 1 1 557 600
-3 838
10 2013 1 1 558 600 -2 753
# ... with 336,766 more rows, and 12 more variables:
sched_arr_time <int>,
# arr_delay <dbl>, carrier <chr>, flight <int>, tailnum <chr>,
# origin <
chr>, dest <chr>, air_time <dbl>, distance <dbl>, hour <dbl>,
# minute <dbl>, time_hour <dttm>
>
```

For the destination airports, in this tibble, you only have the codes of the airports, i.e. IAH, MIA, BQN. Let's see how often which airport was used (I'll explain the group_by() a little later, just pay attention to the result for now):

```
flights %>%
    group_by(dest) %>%
    summarize(Number_of_Flights = n())
```

```
# A tibble: 105 x 2
dest Number_of_Flights
<chr> <int>
1 ABQ 254
2 ACK 265
3 ALB 439
4 ANC 8
5 ATL 17215
6 AUS 2439
7 AVL 275
8 BDL 443
9 BGR 375
10 BHM 297
# ... with 95 more rows
>
```

A tibble: 336,776 × 26

hour <dbl>	minute <dbl>	time_hour <S3: POSIXct>	name <chr>
5	15	2013-01-01 05:00:00	George Bush Intercontinental
5	29	2013-01-01 05:00:00	George Bush Intercontinental
5	40	2013-01-01 05:00:00	Miami Intl
5	45	2013-01-01 05:00:00	NA
6	0	2013-01-01 06:00:00	Hartsfield Jackson Atlanta Intl
5	58	2013-01-01 05:00:00	Chicago Ohare Intl
6	0	2013-01-01 06:00:00	Fort Lauderdale Hollywood Intl
6	0	2013-01-01 06:00:00	Washington Dulles Intl
6	0	2013-01-01 06:00:00	Orlando Intl
6	0	2013-01-01 06:00:00	Chicago Ohare Intl

1-10 of 336,776 rows | 17–20 of 26 columns   1   2   3   4   5   6 … 100 Next

FIGURE 5.7

The result of the left_join() command.

Even though I am very interested in aviation, I do not know every airport. But in the airports data set, the names are written out:

```
> airports
# A tibble: 1,458 x 8
faa name lat lon alt tz dst tzone
<chr> <chr> <dbl> <dbl> <chr> <chr>
1 04G Lansdowne Airport 41.1 -80.6 1044 -5 A America/New...
2 06A Moton Field Munic... 32.5 -85.7 264 -6 A America/Chi...
3 06C Schaumburg Region... 42.0 -88.1 801 -6 A America/Chi...
4 06N Randall Airport 41.4 -74.4 523 -5 A America/New...
5 09J Jekyll Island Air... 31.1 -81.4 11 -5 A America/New...
6 0A9 Elizabethton Muni... 36.4 -82.2 1593 -5 A America/New...
7 0G6 Williams County A... 41.5 -84.5 730 -5 A America/New...
8 0G7 Finger Lakes Regi... 42.9 -76.8 492 -5 A America/New...
9 0P2 Shoestring Aviati... 39.8 -76.6 1000 -5 U America/New...
10 0S9 Jefferson County ... 48.1 -123. 108 -8 A America/Los...
# ... with 1,448 more rows
>
```

Now, how do I get these two data sets together? We can use a join command. I'll run the command once and then explain the details:

```
flights %>%
    left_join(airports, by = c("dest" = "faa"))
```

You can see the result in Figure 5.7: We have joined the two tables, using different variables. In fact, the join commands can also automatically "join" –that is, join the tables – if the same variable names are present in both tables.

However, you also have to be very careful with this, because it does not always achieve what you want. In this example, however, there are no common variable names. The flights data set has a column "Dest"; in the other data set it is called "FAA." Using these two columns, you can link the two data sets by specifying this with by =. Our table now has 26 columns.

The result of this join command is now not stored in the flights record; the next time you enter flights, the record will look the same as before – the result has just been output.

Attention, error source – output of a result

This is a common error: As long as the result of an expression is not written to an object, no change to the data has taken place. The changes are not even in working memory. Write

```
flights %>%
    left_join(airports, by = c("dest" = "faa"))
```

then, the data set flights is not yet changed! Only by writing into an object is the change stored. In the following example, I save into the same object:

```
flights <- flights %>%
    left_join(airports, by = c("dest" = "faa"))
```

However, I strongly recommend that you check the result of a join command before assigning it to an object and proceed exactly as described here. You should also think carefully about whether you assign the result to a new object (then, you will eventually run out of memory with larger data sets, because we will be doing several more joins) or whether you write it to an existing object, for example, flights. The big downside to this is that if you do something wrong, you then have to reload everything again.

Attention, source of error: changes in notebooks

While notebooks are a great thing, the concept also sometimes causes confusion because the order in which code snippets are executed plays a big role. A change in one area of the notebook does not mean that other places in the notebook reflect that change.

Example: If you write everything in flights in this example and change something further up in the code at a later time in the notebook, then it can happen that the code no longer works as you assume because "further up" in the notebook does not mean that the version of flights that the respective code chunk would have produced is then also in memory. flights may have changed in the meantime. Ideally, you should restart R once (via SESSION – RESTART R), empty the environment, and then, run your code again from the beginning.

Many notebooks I receive from students are faulty only because this step was not performed and some objects were in the student's memory that are either not present or modified.

Let's take another look at the join() function. We had used the variant left_join(): why?

- The left_join command takes all the rows of the left data frame (flights in our example) and adds only the rows of the right data frame (airports in our example) where there is a match in the key. If the right data frame does not have a row that matches a row of the left data frame, this row of the left data frame will still be listed.

- The right_join does exactly the opposite of the left_join.

- With inner_join, only those rows are merged where there are matches in both records.

- The full_join() simply joins everything together, regardless of whether there are matches or not.

So you should think very carefully about when to use which join command. Once again, back to the output of our join. In the fourth line, you can see am NA in the column of airport names. This means that there is no value here. How can this be? Let's take a look at all the lines where there is no name but an NA. The function is.na() is already present in Base R:

```
flights %>%
    left_join(airports, by = c("dest" = "faa")) %>%
    filter(is.na(name)) %>%
    select(dest) %>%
    unique()

# A tibble: 4 × 1
dest
```

```
<chr>
1 BQN
2 SJU
3 STT
4 PSE
>
```

The first airport is Rafael Hernandez Airport, which, for whatever reason, has no name here. We can quickly change that with the mutate() function:

```
flights %>%
    left_join(airports, by = c("dest" = "faa")) %>%
    mutate(name = if_else(dest == "BQN",
    "Rafael Hernandez Airport", name))
```

For practice, I'll leave it to you to identify the remaining airport names and adjust them in the data set.

NA, ZERO, and 0

The difference between ZERO and NA: ZERO is first of all not 0. Someone can be 0 years old, i.e. freshly born. NULL, on the other hand, is an object that denotes an undefined value, e.g. because the wrong format was used during a data import. NA, on the other hand, fits, for example, if the age is not known. For example, if the average is calculated, then it would be wrong to simply assume 0 for those where the age is not known. Instead, the NA values are then ignored. In machine learning algorithms, there are different requirements for how NA values are handled.

You have seen before the function group_by(), but I have not explained it. Here is the code again:

```
flights %>%
    group_by(dest) %>%
    summarize(Number_of_Flights = n())
# A tibble: 105 x 2
dest Number_of_Flights
<chr> <int>
1 ABQ 254
2 ACK 265
3 ALB 439
4 ANC 8
```

```
5 ATL 17215
6 AUS 2439
7 AVL 275
 8 BDL 443
9 BGR 375
10 BHM 297
# ... with 95 more rows
>
```

The verb group_by() is immensely useful. We could formulate the preceding expression as follows:

- Show me the flights record,

- and then look for each unique value in the "dest" column,

- how often it occurs.

If group_by() is used in an expression without any other command after it, the effect is not visible. This is a challenge for beginners at first, since the change is only virtual, so to speak. However, something actually happens, but it only becomes clear when the command that follows uses the groups. You can imagine it roughly like this, using an example that everyone can understand:

```
gummy bear bag %>%
    group_by(color) %>%
    summarize(count = n()) %>%
    arrange(desc(number))
```

So from the gummy bear bag, it is first sorted by color, then the gummy bears of each color are counted, and finally the number of gummy bears in a color is shown in descending order.

5.3.5 Regular Expressions and Mutate()

First of all: Many great books have been written about regular expressions (see especially [25]). Regular expressions are about recognizing patterns in strings so that operations can be performed on them. A small example:

```
airports %>%
    mutate(continent = str_remove(tzone, "/.*")) %>%
    mutate(town = str_extract(tzone, "[_a-zA-Z]*$"))
```

We create here with mutate() a new column or two at once. The column "tzone" contains both the continent and the city of an airport. Perhaps, we would prefer to have these separated. Here regular expressions help us immensely. Of course, we could also work with if_else, but that would take quite a long time. Instead, we say in the first regular expression that we want to find everything in the string tzone that starts with the slash so we can remove it. The dot stands for each character and the asterisk for the fact that it can occur one or more times.

In the second mutate() command, we extract what we're interested in, which is everything there is in characters before the line ends. I need to explain this a bit more: With

```
[_a-zA-Z]*
```

all lowercase and uppercase letters are identified, and also the underscore, in any frequency. The dollar sign at the end says that this is the end of the line. That's why this regular expression doesn't find the continent, because the slash doesn't occur in it, so it looks for all strings in which my pattern occurs up to the end of the line. I'll leave it to you to try it out, but before doing so, you should study the list of different operators.

I would like to demonstrate another example of the use of mutate() using the weather table. It is not always necessary to use regular expressions to mutate a cell. In the table, temperatures are given in Fahrenheit, which most Europeans tend to have problems with. The conversion formula is quite simple, and so we can change the temperature without any problems:

```
flights <- flights %>%
    mutate(temp = (temp-32)/1.8)
```

In this example, I'm overwriting the existing table (after I've tested everything, of course), because I'm not going to work with Fahrenheit again.

5.4 Data Visualization

Data visualization is an art, as it involves making properties of the data visible that would otherwise not be easily seen. We will first discuss important visualizations in this section and then ggplot2, a library by – you guessed it – Hadley Wickham.

5.4.1 Data Visualization as Part of the Analysis Process

In an exploratory data analysis, data is visualized to get an overview of distributions and trends, as you have already seen in part in the previous section. You should not underestimate the importance of visualization in exploratory data analysis. This is demonstrated by the Anscombe quartet in a particularly impressive way: it consists of four sets of data points that have almost identical statistical measures [8]. The x-values all have the same mean and

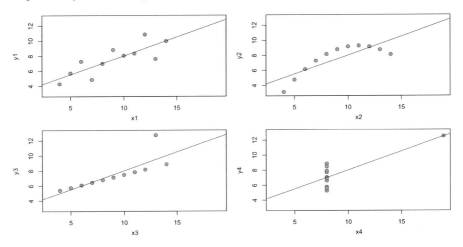

FIGURE 5.8
The data points of the Anscombe quartet visualized.

variance, and for the y-values, these are the same except for two and three digits, respectively. Also the correlation coefficients and the linear regression (see Chapter 6, "Forecasting") are also the same except for a few digits.

However, when these data points are plotted, it becomes clear that they differ greatly, as can be seen in Figure 5.8. The Anscombe quartet should serve as a warning that a plot is not an option but a fundamental part of the analysis. Plots help the user to understand the data that are available.

5.4.2 Data Visualization as Part of the Reporting

Visualizations are often used in the reporting of a project. First of all, it is obviously important to choose the technically correct visualization. Just because, for example, Excel suggests a bar chart does not mean that this is the best visualization of a data set. A good bad example is the pie chart from the old version of Google Analytics, in Figure 5.9 to see. A pie chart is only useful when we want to see the ratios of two or more groups to each other. One is either in one group or the other, but not in more than one. However, this is exactly what can happen in this context, because one can be on a website for the first time within the observation period, i.e. a new visitor, as well as having come back again afterwards, i.e. a returning visitor at the same time. The pie chart, as shown in Figure 5.9 is in any case the wrong choice (see also Alby, 2019 [5]).

In addition, it is necessary to consider how information is presented in a way that makes it easier for recipients to absorb it. The information value of a report must be enhanced by data visualization. The inclusion of a graphic must not be an end in itself. That has not been done in this example either

■ New Visitor   ■ Returning Visitor

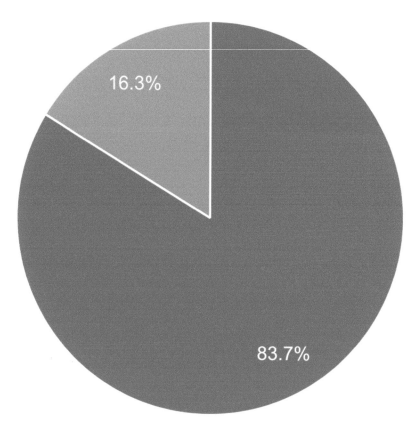

FIGURE 5.9
A pie chart that should never have become one, in Google Analytics.

So, while data visualization during analysis and modeling is primarily about seeing properties of the data points, in a report the focus is on highlighting the information to be communicated. Where appropriate, data visualizations from the analysis phase can be used, but it is important to verify that these graphics are actually appropriate for those who will consume them. An ROC AUC curve is certainly appropriate for experts, but if the consumer is not a Data Scientist, he or she must first be introduced to the concept in an elaborate way. Communication, as described in Chapter 3, "Anatomy of a Data Science Project," is an important component in Data Science, and so is the creation of data visualizations. Or, as John Tukey put it:

The greatest value of a picture is when it forces us to notice what we never expected to see.

It is helpful to be aware of the intent of a report. Is a data visualization intended to represent something positive? Is it to suggest a corrective action based on an unfavorable development? However, this does not mean that you should create data visualizations in such a way that manipulation occurs. While the trained eye is quick to identify such a manipulation, the less trained eye is quickly led by the nose. It is necessary to do everything possible to optimize the credibility of data collection and presentation. [18] summarizes the task well:

1. Understand the context: Who is being communicated to, what is the person supposed to do (or not do), how is the person being communicated with, and what is the "flow" for the story?

2. Choose a suitable visual representation (this is what the next sections are about)

3. Remove unnecessary information

4. Ensure that the reader can focus on something

5. Think like a designer while telling stories (see Chapter 3).

The recipient is in the foreground here; everything is aligned with him. Only when the first task has been solved, consideration is given to which visual representation is the right one.

5.4.3 Plots in Base R

You have already seen the simplest kind of plot in the notebooks, it simply consists of plot(cars). This form of visualization is called a *scatter plot*. In order for something to be plotted on two axes, there must be values for both x and y in the data set, which is the case in the cars example. The application of plot(), which becomes visible when the commands are completed, looks like this:

```
plot(x, y, ...)
```

First comes the variable x, then the variable y. Optionally, you can specify further parameters. For example, with the parameter type = "l" you exchange the points for a line.

plot() automatically takes the first variable from the data set as the value for the x-axis and the second for the y-axis. If a data frame has more than two variables, you must select the variables to be plotted. As an example, we want to examine two variables from the data set mtcars in more detail. The data set contains the number of cylinders and the consumption in miles per gallon for each car.

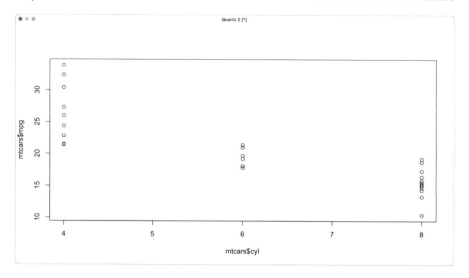

FIGURE 5.10
A simple plot of the number of cylinders and miles per gallon from "mtcars".

```
plot(mtcars$cyl,mtcars$mpg)
```

The resulting plot in Figure 5.10 has some issues. For example, the labels for the x and y axes are not necessarily immediately understandable to outsiders. A title in the plot would also be useful.

You can achieve this with the following parameters:

```
plot(mtcars$cyl, mtcars$mpg, xlab="number of cylinders",
ylab="miles per gallon", main="ratio number of cylinders
to miles per gallon")
```

xlab is for the x-axis label, ylab for the y-axis, main for the title. You can examine the result in Figure 5.11. However, this visualization is still not really useful.

On closer inspection, the number of cylinders is a number, but actually the number in this context is more of a category. In this data set, cars exist with 4, 6, or 8 cylinders. However, the plot also contains markers for 5 and 7 cylinders, which are probably rare in cars. By adding the xaxp parameter, you can specify where the tick marks are are visible. Here, a starting value (4), the last value (8), and the interval (2) where they are visible are specified first.

```
plot(mtcars$cyl, mtcars$mpg, xlab="number of cylinders",
ylab="miles per gallon", main="ratio of number of
cylinders to miles per gallon", xaxp = c(4,8,2))
```

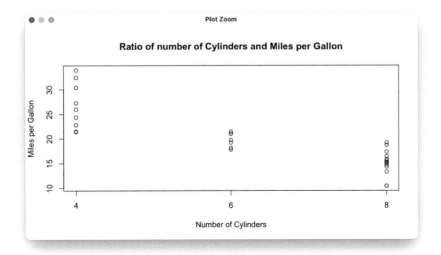

FIGURE 5.11
The previous plot with a proper labeling.

One way to plot the relationship between these two variables is a so-called boxplot, sometimes also called a whisker plot. The code for generating a boxplot is slightly different than for a normal plot:

```
boxplot(mtcars$mpg~mtcars$cyl, xlab="number of cylinders",
ylab="miles per gallon",
main="ratio number of cylinders to miles per gallon")
```

The tilde $\sim$ between the two variables gives a ratio of the variables to each other, you could express it as "the miles per gallon conditioned on the number of cylinders." You can see the result in Figure 5.12.

A boxplot needs some introduction if you are dealing with it for the first time. In such a plot, numerical data are represented by their quartiles. Quartiles are used to divide numerical data into 4 equal parts, starting with the smallest number and then sorting in ascending order to the largest number. The first quartile starts with the smallest number and goes up to half of the data to the median. The box, which is the gray box, contains 50% of the data points, which is the second and third quartile data. The black line in the gray box is the median..

In Figure 5.13, the lower quartile can be seen to the left of the box and the upper quartile to the right of the box. The lines outside the gray box, the "antennas" or whiskers contain in most boxplots the 1.5-fold interquartile range, that is, what corresponds to the length of the box. Anything beyond this is interpreted as an outlier. However, the whisker is only set to 1.5 times the interquartile range if there are values there. In the example, there are no

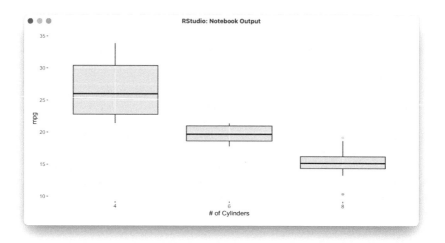

FIGURE 5.12
A boxplot with the data from "mtcars".

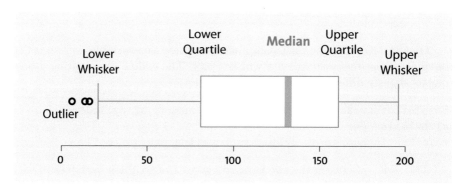

FIGURE 5.13
A boxplot and its elements.

data values beyond the upper whisker, so that it is plotted only up to that point.

A boxplot shows not only the distribution, but also an interpretation of where outliers lie. However, the definition that anything beyond 1.5 times the interquartile range is an outlier is not set. After all, something does not always have to be an outlier just because it is outside that distance.

Another chart frequently used in analysis is the histogram. It is a visualization of a frequency table. Here, the values are divided into classes, so that each class is not listed individually.

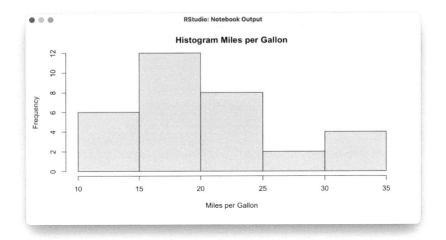

FIGURE 5.14
Histogram with standard classes.

```
hist(mtcars$mpg, xlab="Miles per Gallon", ylab="Frequency",
main="Histogram Miles per Gallon")
```

In Figure 5.14, you can see the result of this command. R chooses itself the number of classes (in English bins), which is not always advantageous.

However, in the command, you can also specify a parameter that specifies the number of classes. Strangely enough, it is not called bins, but breaks:

```
hist(mtcars$mpg, xlab="Miles per Gallon", ylab="Frequency",
main="Histogram Miles per Gallon", breaks=10)
```

In Figure 5.15, the distribution changes slightly. Four cars get far more miles per gallon than the rest. However, it is also obvious that the hist()-command has not implemented the exact specification of the classes as specified. In fact, the specification is declared as a recommendation in the help accompanying the command; R tries to adhere to it, but does not necessarily do so.

However, the classes can also be specified in advance, for example

```
hist(mtcars$mpg, xlab="Miles per Gallon", ylab="Frequency",
main="Histogram Miles per Gallon",
breaks=c(10,12,14,16,18,20,25,28,30,35))
```

The classes don't even have to be the same size. However, brainpower must be invested here in advance as to which classes make sense. If the classes are too granular, then the histogram may no longer make sense, because data

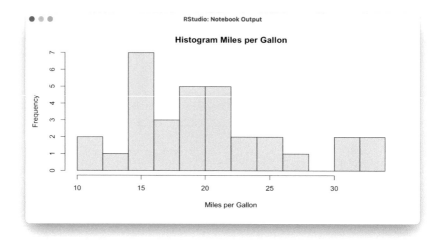

FIGURE 5.15
Histogram with "breaks=10" but more than 10 classes.

points are no longer summarized. If the classes are too large, then the patterns in the data may no longer be obvious.

You can discover real surprises in histograms, for example, that a distribution with one mode in the standard histogram actually has more than one mode when the bins are defined differently. It is definitely worthwhile to "play" with the classes. By the way, a histogram is also a plot, even though the term "plot" does not appear in the command. Data are plotted, no matter whether in a histogram with the hist() function or in a scatter plot with the plot() function, with the plot() command.

5.4.4 ggplot2: A Grammar of Graphics

The previous examples show that creating graphs is not necessarily difficult, but the functions work differently in some cases. plot() would like to have a comma between x and y, while boxplot() would like a tilde. To simplify this, ggplot2 was born. ggplot2 has become a popular package for data visualization, designed following the Grammar of Graphics by Leland Wilkinson. The idea is that a grammar provides rules to describe in a short and concise way the components of a graph. Wickham's approach adds the idea that such a grammar can be implemented with layers, where each layer defines a component of a graphic. Such layers could be:

- One record

- The aesthetic assignment of which data point belongs to what

- Scale values

- Geometric objects (points, lines, etc.)

- Additional plots within the basic plot

- A coordinate system

- Additional statistical information, such as the median in a boxplot or a confidence interval

So instead of knowing for each command what the respective parameters are called and how they have to be set, the Layered Grammar of Graphics approach allows composing a graphic with multiple layers. I will illustrate this with a simple example. First, the library is loaded:

```
library(ggplot2)
```

The consumption in connection with the number of cylinders is to be visualized again.

```
ggplot(mtcars, aes(x=as.factor(cyl), y=mpg)) +
    geom_boxplot(fill="slateblue", alpha=0.2) +
    xlab("number of cylinders")
```

The first line defines the data set and the aesthetic assignment, the second the geometric shape with additional parameters, and the third a scale label. As with dplyr, the individual lines build on each other. This code results in the plot in Figure 5.16. Apart from the fact that the whiskers look different here than in the Base R plot, there is an additional grid in the background, which, of course, can be modified.

Somewhat unusual in ggplot2 is that, at least still in 2023, a different symbol is used at the end of each line than in dplyr. This leads to somewhat confusing code combinations, because ggplot2 can be used directly in dplyr expressions:

```
mtcars %>%
    select(cyl, mpg) %>%
    ggplot(., aes(x=as.factor(cyl), y=mpg)) +
    geom_boxplot(fill="slateblue", alpha=0.2) +
    xlab("number of cylinders")
```

However, the combination of different symbols at the end of each line can also be justified by the fact that the $\% > \%$ in dplyr has a different function than the $+$ in ggplot2: while in dplyr the output of one line means the input of the next line, in ggplot2 it simply "puts a layer on top".

In the following chapters, I will use ggplot2 more often and explain more examples.

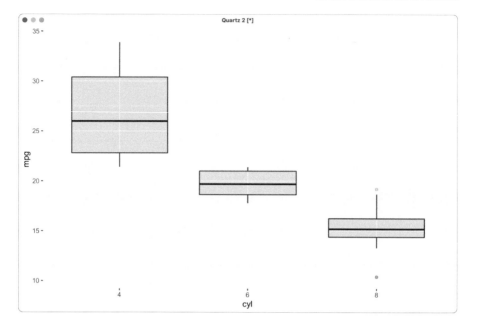

FIGURE 5.16
A boxplot created with ggplot2.

5.5   Data Analysis

You have now learned almost everything you need for exploratory data analysis – notebooks, the Tidyverse, and data visualization. To dive deeper, you'll still need some basic statistics knowledge. While I can't give you a complete basic course here, there are a few concepts we should start looking at. For an always up-to-date list of recommendations for getting started with statistics with R, please check out my website at https://alby.link/statistik. There you will also find an online basic course.

We're now taking a closer look at the nycflights13 data set, and what we want to know is what factors actually cause flights to be delayed.

First of all, we merge all the tables:

```
my_flights <- flights %>%
    left_join(airlines) %>%
    rename(airline = name) %>%
    left_join (airports, c("dest" = "faa")) %>%
    rename(airport = name) %>%
    left_join(weather) %>%
    left_join(planes, by = "tailnum")
```

We get 44 columns, and the number of rows should remain the same. If not, then something has gone wrong. The other necessary transformations can be found in the notebook on the website. First, we look at the distribution of delays. Then, we look at one variable before we look at the relationship of the variables to each other in the next step. For the distribution we first use the function summary():

```
> summary(my_flights$dep_delay)
   Min. 1st Qu. Median Mean 3rd Qu. Max.
11.00 20.00 38.00 58.53 74.00 1301.00
>
```

We can already see here that we have a right-skewed distribution, because the arithmetic mean (which is commonly called the *average*) is above the median (the value that lies exactly in the middle when you line up all the delays sorted).

Median, middle income, what's the difference?

When the news talks about a middle income in Germany, it usually means the median and not an average income. The distribution of incomes looks similar to the distribution of delays: Few earn a lot, many earn less. If average income were actually communicated in the statistics, then the majority earns below average, and that certainly doesn't feel good to this majority. With the median income, the effect is different, because still many people could be exactly on the median, so fewer than half are below the median. The median is more robust against outliers, and that is used here.

We plot the distribution once in a histogram:

```
data <- my_flights %>%
    filter(dep_delay > 10)
    ggplot(data = data, aes(x=dep_delay)) +
    geom_histogram(binwidth = 15)
```

The result can be found in Figure 5.17. As expected, you can see that the distribution is curved to the left. Obviously, large delays are rather the exception. This is already reassuring.

Another way to look at the distribution, as discussed in the previous section, is the boxplot:

```
ggplot(data = data, aes(x=dep_delay)) +
    geom_boxplot() +
    scale_x_continuous(breaks= seq(0, 1350, by=60))
```

In the plot in Figure 5.18, you can see that 50% of all delays are between 20 and 74 minutes (as already output by summary()), but the advantage of this plot is that we can also see the large outliers. In the histogram, this was not possible, because a single point is no longer displayed if the frequencies on the y-axis are not in the same range.

Next, we visualize the relationships among the variables with a pairs plot which is a scatter plot of several variables, so to speak:

```
pairs(~dep_delay + speed + distance + arr_delay + engines + seats +
temp + wind_speed, data=my_flights)
```

Your computer will now have a bit more to do, but after a few minutes you should have a plot that looks something like the one in Figure 5.19. I didn't select all variables here, as that doesn't make sense either; more about that in a moment.

We see in the plot a clear relationship between the delay in departure and the delay in arrival. This is not surprising, because if a plane takes off later, it will land later. Let's take a closer look by considering all flights that had a departure delay of more than 10 minutes:

```
my_flights %>%
    select(dep_delay,arr_delay) %>%
    filter(dep_delay > 10) %>%
    plot()
```

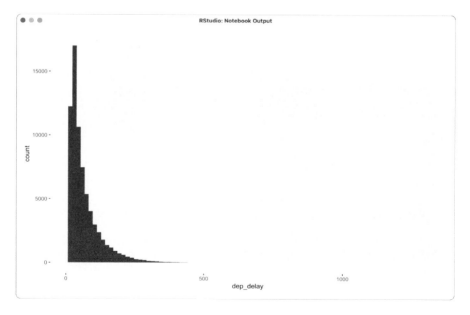

FIGURE 5.17
Histogram of delays of 10 minutes and more.

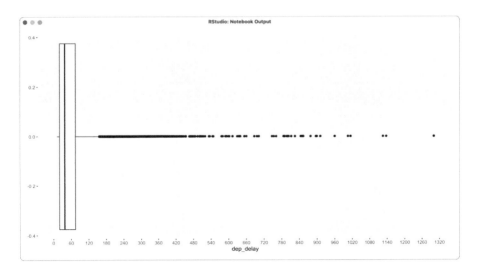

FIGURE 5.18
Boxplot of delays of 10 minutes and more.

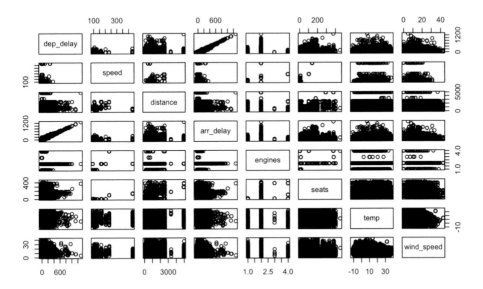

FIGURE 5.19
The output of pairs().

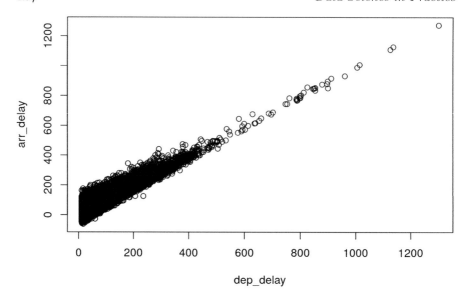

FIGURE 5.20
Even those who depart reasonably on time can still be quite a bit late.

You can see the result in Figure 5.20. What is interesting here is that even flights with little departure delay can still arrive quite a bit later, but also, flights that started later can arrive earlier.

We additionally perform a correlation test. A correlation is a statistical correlation, which does not necessarily mean that a correlation actually exists. Also, the direction is usually not known. In our example, however, we already know that there is a real correlation, and we also know the direction.

First of all, we store the data so that we can use it:

```
> dep_arr_delay <- my_flights %>%
+ select(dep_delay,arr_delay) %>%
+ filter(dep_delay > 10)
```

The correlation test:

```
> cor.test(dep_arr_delay$dep_delay,dep_arr_delay$arr_delay)

        Pearson's product-moment correlation
data: dep_arr_delay$dep_delay and dep_arr_delay$arr_delay
t = 805.85, df = 82299, p-value < 2.2e-16
alternative hypothesis: true correlation is not equal to 0
95 percent confidence interval:
0.9413109 0.9428479
sample estimates:
```

```
cor 0.
9420843
```

We obtain a correlation coefficient of over 0.94, which means a strong correlation. This is not surprising, but it also does not explain what causes departure delays.

More about Correlation

For the sake of completeness: Of course, not only is the correlation coefficient important to look at, but also the other values. Especially the p-value is important here (in this case 2.2e-16, i.e. 0.000000000000022. Many books have been written about the p-value, but still it is difficult to explain. A very simple explanation: If we assume that the departure delay and the arrival delay would have nothing to do with each other, how likely is it that we would see these values? The lower the value, the lower the probability. A significance level of 0.05 is often taken as the limit above which the null hypothesis (there is no statistical relationship) is rejected and the alternative hypothesis (there is a statistical relationship) is accepted. The significance level depends on the research question. If there would be serious consequences if the null hypothesis was wrongly rejected, then a lower level such as 0.01 should be chosen.

We cannot always simply use cor.test() in every case. Strictly speaking, we should not have used Pearson's product-moment correlation, because our variables are not normally distributed. In this case, we should have used Spearman's correlation (you can specify it with *method = "spearman"*).

It gets even more complicated if, for example, we want to measure the correlation between departure delay and a categorical variable. If in doubt, ask a statistician you trust.

If we look at the other plots in Figure 5.19, there are hardly any patterns to be seen. Only in the number of machines, there might be something, and we look at that again in more detail.

```
engines_delay <- my_flights %>%
    select(engines,dep_delay) %>%
    filter(!is.na(dep_delay)) %>%
    filter(!is.na(engines))
    boxplot(engines_delay$dep_delay~engines_delay$engines,
        varwidth=TRUE)
```

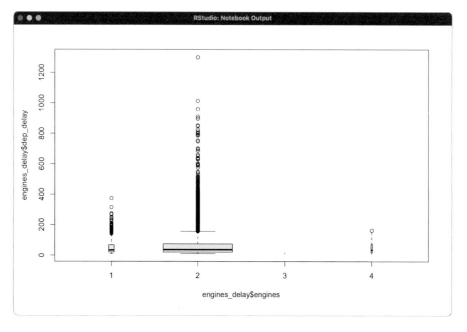

FIGURE 5.21
A boxplot of delays as a function of the number of engines.

For the parameters of the boxplot() function, we use varwidth=TRUE, which allows us to include the number of data in the visualization. The wider the box, the more data there is. The result in Figure 5.21 shows that there seem to be only a few flights with aircrafts with three or four engines.

One could now interpret that the presence of two engines is most likely to cause delays. This is, of course, humbug, because if most flights take place with twin-engine aircraft, then the probability of being late is also much greater.

```
> my_flights %>%
    + group_by(engines) %>%
    + summarize(number_engines = n())
# A tibble: 5 × 2
engines number_engines
<int> <int>
1 1 2014
2 2 282005
3 3 7
4 144
5 NA 52606
>
```

However, it is also interesting to note here that we have over 50,000 rows

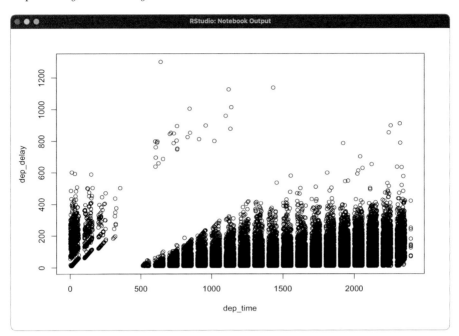

FIGURE 5.22
Does departure time affect delay?

where no engine count is given. Apparently, we do not have all aircraft types in our data set.

Before we give up, let's look at one more data point that is missing from our scatter plot. Does the time of day have anything to do with whether an airplane takes off later?

```
my_flights %>%
    select(dep_time,dep_delay) %>%
    filter(dep_delay>10) %>%
    plot()
```

You can see the result in Figure 5.22. Delays build up slowly in the first hours of the day. However, starting at 6 a.m., some planes take off much later than scheduled, with delays of more than 600 minutes, i.e., at least 10 hours. We have two different types of delays starting at 6 a.m.: very small ones up to one hour, and very large ones starting at 10 hours. These flights with a long delay should have started the previous day, so they have less to do with the delays of the flights of that day.

It is also surprising that there are actually no on-time departures in the early morning hours; the further after midnight it gets, the later the flights leave, although the long delays do not increase. Between 4 a.m. and 5 a.m., no flights seem to take off at all.

By the way, you might also notice the spaces between the times. It seems as if there is a pause between the hours. However, this is not the case. Since the time is taken as an integer between 500 and 559, for the time between 5:00 and 5:59, we automatically have spaces, because the time between 5:60 and 5:99 does not exist, but plot() assumes it does.

Obviously, the departure delays are caused by factors that we cannot easily find in the data. A plane cannot depart until it has arrived beforehand, but we do not know beforehand whether the plane has actually arrived on time. By having a break in the early morning hours, departure delays caused by late arriving aircraft can be mitigated somewhat. The plot in Figure 5.22 could also suggest that increased late morning operations are causing delays.

Many more exciting questions can be answered with this data set. You can find an example of an analysis on the website for the book.

6

Forecasting

How many new customers are expected for how much advertising budget? What is a good price for a used SLR camera? Are the changes in the number of visitors to our website what we would expect? These questions can be answered with simple models.

In this chapter, we look at two different approaches. On the one hand, we will look at regression analyses, where the goal is to find out whether we can predict the impact of a value x on a value y. Example: I want to buy a used SLR camera. The higher the shutter count, the lower the value, because at some point, the shutter has to be replaced. The sellers of used cameras know this and usually price it in. So, if we have the data from many used cameras, including price and the shutter count, can we form a model that will give us a fair price when the shutter count is known?

On the other hand, we deal with time series and forecasts based on them. Let's say you have a website that is frequently visited on weekdays during regular business hours, but hardly ever in the evenings and on weekends. You can see a clear pattern here and make predictions based on how many visitors you will have at certain times. But, what if you suddenly had an incredible number of visitors on the weekend? You can identify such outliers automatically with various forecasting algorithms.

6.1   Linear Regression

Before we discuss linear regressions, we will recall what linear functions are. The example that all my students understand is a cell phone contract. I pay 14.90 dollars every month for my contract, sort of as a base price. For every gigabyte I use, I pay 50 cents. My costs are variable, but I can calculate them relatively easily:

$$total\_price = number\_of\_GB\_used \cdot price\_per\_GB + basic\_price$$

In mathematical terms, it looks like this:

$$f(x) = m \cdot x + b$$

DOI: 10.1201/9781003426363-6

m is the *slope* and *b* is the *constant*. *x* is the number of gigabytes I used.

In a linear regression, we try to find out the values of this function by passing existing data to the algorithm. Most of the time the data points are not exactly on the straight line; otherwise, we could simply plot them in a coordinate cross and read the formula.

6.1.1   How the Algorithm Works

Linear regression is, so to speak, the "Hello World" of machine learning and one of the oldest methods of machine learning, which found its origin in the considerations of Legendre and Gauss at the beginning of the 19th century. In fact, some basic principles of many algorithms can be recognized in linear regression, so you should at least skim this chapter even if you are not necessarily interested in linear regression.

Linear regression is an attempt to explain a *dependent variable* by one or more *independent variables*. In contrast to correlation, where the aim is to examine how strong a relationship is (without being able to establish causality), regression is concerned with the best possible prediction of the dependent variable. Example: The more marketing budget spent, the more prospects can be attracted to a product (*leads* in marketing parlance). Such a relationship is shown in the scatter plot in Figure 6.1.

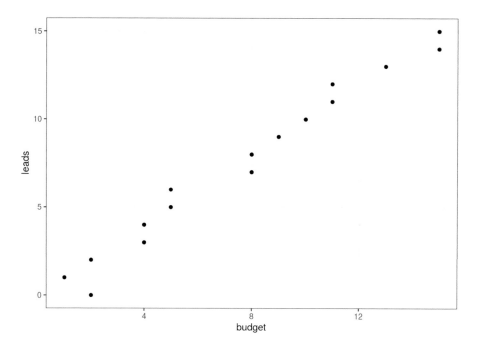

FIGURE 6.1
Scatter plot of budget to leads ratio.

You could now draw such a line through the points with a ruler. Is this the line that leads through these points in the best possible way? In this context, "best possible" means that all points have the smallest possible distance to the line. Very likely, we will not find this line so easily by eye; the line might have a little more, or maybe a little less, slope.

To find out which line goes best through the points, we first perform a small thought experiment. Not only is one line drawn in the plot, but quite a few, and for each line, we calculate how far the points that are not exactly on the line are from this line. Of course, these lines are not drawn all at once, but one after the other. And for each calculation, we look at how far the points are from the respective line. In Figure 6.2 we see such a line.

Sometimes, these distances to the line are called *errors* but, of course, they are not errors in the usual sense; they are simply the distances that the particular line produces from the given points. Another term for this is *residuals*. The graphic in Figure 6.3 illustrates the approach with the line drawn before.

The distances of the points to the line are squared (to get rid of negative numbers) and then added to each other. For each line, a measure is created, the sum of the squared deviations that describes the distance of all points to the respective line. Since each line produces a different sum of squared

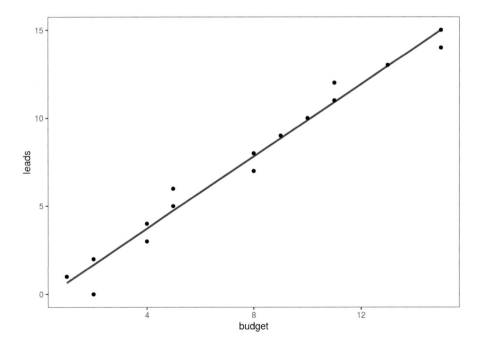

FIGURE 6.2
Regression line through the data points.

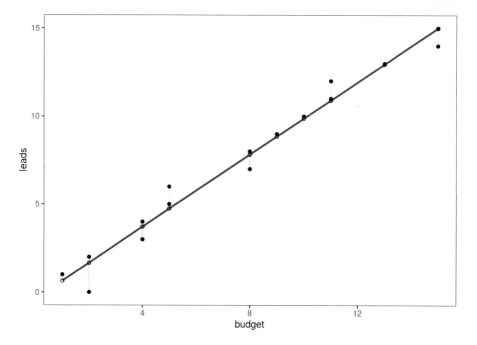

FIGURE 6.3
The same straight line as in the previous figure, here with distances to the data points.

deviations, we can now select the line that produces the lowest sum. This is then the optimal line.

Measure and minimize errors, distances, and the like

The principle used here, that distances are measured or errors are minimized, is fundamental to many machine learning algorithms. In the next chapters you will come across algorithms again and again, so you should definitely understand this concept for further understanding.

In the plot in Figure 6.3, you can observe a prerequisite for a linear regression, namely, that the variances of the residuals are equal. This is called *homoskedasticity*. If this is not the case, we speak of *heteroskedasticity*. Heteroskedasticity is what variance heterogeneity is for regression analysis. Put simply, if the residuals have different distances from the line, that is, their variance is heterogeneous, then we speak of heteroscedasticity. If this is present, the resulting model must be taken with a lot of caution, because some points are further away from the regression line than others.

6.1.2 Linear Regression in R

We create the first regression analysis once for general overview in only six lines of code and then go through the details. The code can be found on the book page https://alby.link/datascience and in the download area. First, we create a data set:

```
leads <- c(0,1,2,3,4,5,6,7,8,9,10,11,12,13,14,15)
budget <- c(2,1,2,4,4,5,5,8,8,9,10,11,11,13,15,15)
work_hard <- data.frame(budget,leads)
```

Now, we create the model and plot it:

```
fit <- lm(work_hard$leads ~ work_hard$budget)
plot(work_hard)
abline(fit)
```

For the examples so far, we have used the ggplot2 package, so the output of this plot looks slightly different than that of the Base R plot. Let's look at the code in detail. We assign the result of the regression analysis, which we assign to the fit object by calling lm() (for linear model).

```
fit <- lm(work_hard$leads ~ work_hard$budget)
```

As parameters, we first pass the dependent variable leads to the function lm() and then, separated by a tilde ∼, the independent variable budget. This order is immensely important! Consider for a moment what happens if you swap the dependent and independent variables. A little more about this in a few lines.

With plot(work_hard), you plot your data, but we need a second line:

```
abline(fit)
```

If you run plot(work_hard) in a notebook first and then abline(fit), you will get an error message:

```
Error in int_abline(a = a, b = b, h = h, v = v,
untf = untf, ...) : plot.new has not been called yet
```

If you run the code in the console, or together in a code snippet, then you won't have this problem. abline() refers to the current plot that was created in advance, and needs just that context, which is lost in a notebook for whatever reason if you run the lines individually.

But, it goes further, because abline() creates a straight line. The function abline() gets as parameter the object to which we assigned the result of our linear regression. What's actually in there? Just enter fit:

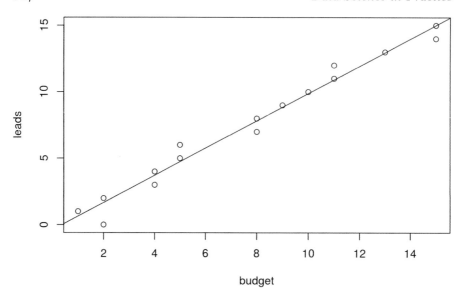

FIGURE 6.4
A regression visualization created with plot() and abline().

```
> fit

  Call:
lm(formula = work_hard$leads ~ work_hard$budget)
Coefficients:
(Intercept) work_hard$budget
-0.384 1.026
```

We see the call and two coefficient values, namely the intercept, which stands for the intercept on the y-axis, and another value for work_hard$budget. Let's think back to Section 6.1, where we were talking about linear functions at the beginning. Then 1.026 is our slope. The only thing missing is the constant. We get it indirectly by drawing our line at −0.384 on the y-axis in our head and then calculating the slope for each x. The line is ready. The line is ready.

However, we can immediately see a problem here: Just as there cannot be a negative budget, I cannot, for example, pay an amount of 0 dollars in my cell phone contract mentioned at the beginning. The intercept should therefore be treated with a great deal of caution.

But there is another aspect of abline() that should be mentioned here. Assuming the x and y values were swapped in the data frame, the plot would look like the plot in Figure 6.5. At first sight, this would not look so bad, but it is the right straight line in the wrong plot. So it is immensely important that you pay attention to the correct order in which the variables are plotted. Conversely, as described earlier, the other possibility for error is that you

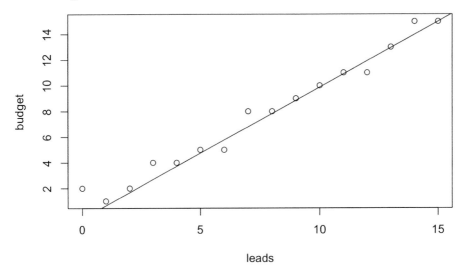

FIGURE 6.5
Right straight line, wrong plot.

have the correct plot but the wrong straight line because you have swapped the dependent and independent variables.

Another source of error, though more in the interpretation and less in the code, is the search for the intercept. The Faithful dataset, which I will discuss in more detail in Chapter 10, "Workflows and Tools," serves as an example:

```
lm(formula = faithful$eruptions ~ faithful$waiting)

  Coefficients:
(Intercept) faithful$waiting
-1.87402 0.07563
```

The intercept is -1.87402, but if we plot the data and the model, it doesn't quite make sense:

Only at second glance does it become clear that the x-axis does not start at 0 and therefore, the intercept seems to be wrong at first.

6.1.3 Interpretation of the Results

Let's take a closer look at the model's data again, this time using summary(fit) to output the details:

```
summary(fit)
## ##
Call:
```

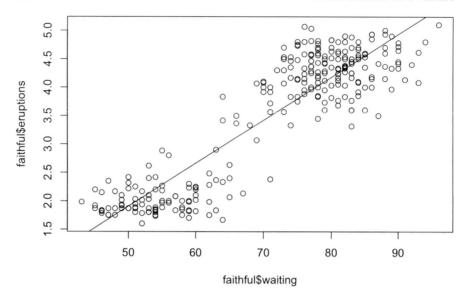

FIGURE 6.6
Here it is worth taking a second look at the x-axis and where it starts.

```
## lm(formula = work_hard$leads ~ work_hard$budget)
## ##
Residuals: ##
Min 1Q Median 3Q Max ##
-1.6671 -0.1791 0.1412 0.2945 1.2562
## ##
Coefficients: ##
Estimate Std. Error t value Pr(>|t|)
## (Intercept) -0.38399 0.38328 -1.002 0.333
## work_hard$budget 1.02556 0.04317 23.754 1.03e-12 *** ##
--- ##
Signif. codes: 0 '***' 0.001 '**' 0.01 '*' 0.05 '.' 0.1 ' ' 1
##
## Residual standard error: 0.7668 on 14 degrees of freedom
## Multiple R-squared: 0.9758, Adjusted R-squared: 0.9741
## F-statistic: 564.3 on 1 and 14 DF, p-value: 1.032e-12
```

The output is (even though it is a "summary") a lot more detailed compared with the previous command. The output contains more important information:

- For the residuals, a reasonably uniform distribution is desirable. What does this mean exactly? In the example, the median is somewhat closer to the third quartile than to the first quartile, the minimum value is somewhat further away from the median than the maximum value. The residuals are therefore not evenly distributed around the straight line, or not exactly normally distributed. The normal distribution of the residuals is a prerequisite for drawing conclusions from a linear regression. In this example, we are at least close to the normal distribution.

- The coefficients I have already explained; here, we now get more values on the standard error, the t-values as well as the p-value. The null hypothesis is that the respective independent variable (in this case, we have only one, the budget) does not correlate with the dependent variable. If the p-value is below alpha, then this null hypothesis is rejected. In the preceding case, this is exactly what happens; however, we have a p-value above alpha for the constant. However, this is not a problem; the intercept is also often ignored because it does not make sense, as already described.

- R-squared is a measure intended to represent how well a model fits the data. Its value is between 0 and 1; the closer the value is to 0, the less the regression can explain the variance of the dependent variable. In the preceding example, over 95% of the variance can be explained by the independent variables. The more variables there are, the more R-squared will increase, so for data sets with many variables, an adjusted version of R-squared is used, simply called Adjusted R-squared.

- Finally, we get the values of the F-statistic.

The assessment of these values is not optional, but is part of the mandatory program.

6.1.4 Advantages and Disadvantages

Even if the algorithm is relatively simple, a lot can still be done wrong. It starts with the question of whether the data is suitable for linear regression in the first place. After all, just because a line fits through data points, this does not mean that the line is the right solution to the problem.

Figure 6.7 shows a commonly used example where linear regression gets wonderful values (p-value below 0.05, R-squared at 0.99!), but still, it is not a viable option here. Instead, logistic regression would be useful. Examples of classifications using linear regression can also be found in the literature. For example, Foreman (2014)[26] created a classification using linear regression in Excel to identify whether customers are pregnant based on shopping baskets, even though they may not know it themselves yet. However, I would not recommend a classification with a linear regression, as explained by the example in Figure 6.7.

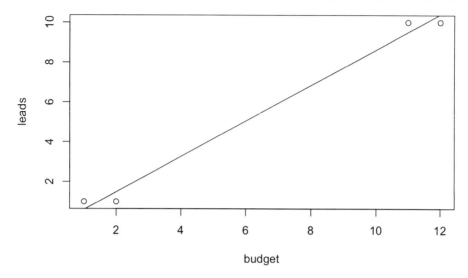

FIGURE 6.7
Is linear regression a good option here?

6.1.5   Non-Linear Regression

Most Data Science introductions start with linear regression and then skip over non-linear regression. This is probably because we can't just throw an algorithm at the data; we have to make estimates of the parameters. This is because in non-linear regression, there is no line going through data points, but a curve. Only after bringing in estimates for the parameters of the curve can we approach it with nls(). The acronym stands for Nonlinear Least Squares. We have already learned about the principle of least squares.

First, we create a data set and plot it:

```
set.seed(55)
x<-seq(0,50,1)
y<-((runif(1,10,20)*x)/(runif(1,0,10)+x))+rnorm(51,0,1)
plot(x,y)
```

You can see the plot for this in Figure 6.8. Quite obviously, linear regression makes little sense here. We test for correlation:

```
> cor.test(x,y)

        Pearson's product-moment correlation
data: x and y
t = 5.6341, df = 49, p-value = 8.526e-07
alternative hypothesis: true correlation is not equal to 0
95 percent confidence interval:
```

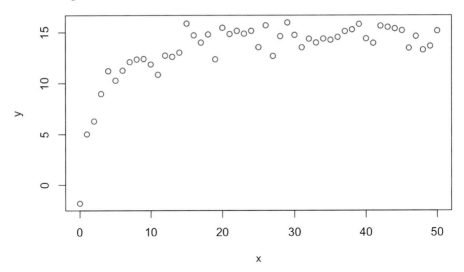

FIGURE 6.8
No matter how much you squint, a straight line doesn't really fit in here very well.

```
0.4248315 0.7696081
sample estimates:
      cor
0.6270073
```

There is a moderate correlation, but the regression line does not seem to fit (see Figure 6.9):

```
fit2 <- lm(y~x)
plot(x,y)
abline(fit2, col = "red")
```

This is also evident in the details of the model:

```
> summary(fit2)

  Call:
lm(formula = y ~ x)
Residuals:
Min 1Q Median 3Q Max
-11.7681 -0.6155 0.2952 1.2522 3.9438
Coefficients:
Estimate Std. Error t value Pr(>|t|)
(Intercept) 9.95572 0.67582 14.731 < 2e-16 ***
x 0.13125 0.02329 5.
```

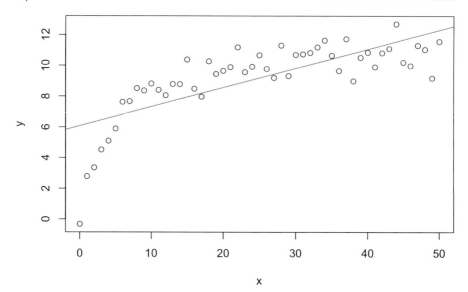

FIGURE 6.9
Regression line through non-linear data points.

```
634 8.53e-07 ***
---
Signif. codes: 0 '***' 0.001 '**' 0.01 '*' 0.05 '.' 0.1 ' ' 1
Residual standard error: 2.449 on 49 degrees of freedom
Multiple R-squared: 0.3931, Adjusted R-squared: 0.3808
F-statistic: 31.74 on 1 and 49 DF, p-value: 8.526e-07
```

The residuals are not normally distributed because the minimum is -11.7681, far from the rest, and our R-squared value is less than 0.4. Obviously, we are not getting anywhere like this.

As mentioned at the beginning, we can only work with nls() for a non-linear regression analysis once we have an estimate of the parameters. This requires some practice. For the preceding data, the following function is estimated:

```
> m<-nls(y~a*x/(b+x))
```

In fact, the data are more correlated here:

```
cor(y,predict(m))
## [1] 0.9496598
```

And, our plot looks better now, as you can see in Figure 6.10:

```
plot(x,y)
lines(x,predict(m),lty=2,col="red",lwd=3)
```

The big difficulty here, however, is estimating the function.

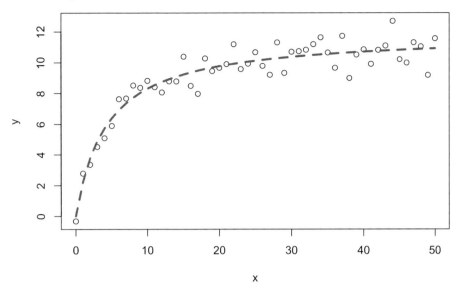

FIGURE 6.10
Regression with nls().

6.1.6   Small Hack: Linear Regression with Non-Linear Data

You should use this hack with a lot of caution. It may be possible, depending on the curve, to perform a linear regression with a data transformation, even if the data does not allow it. To do this, the data is logarithmized.

What is a logarithm?

The best way to explain a logarithm is by using an exponent:

$$a = b^x$$

In a numerical example:

$$9 = 3^2$$

But what if the exponent is known, but not the base? This is exactly what the logarithm solves:

$$x = log_2 a$$

With numbers:

$$3 = log_2 9$$

Another example:

$$5 = log_2 25$$

It is noticeable that the numbers to the left of the equal sign do not grow as fast as the number to the right. And that's exactly what we use for our little hack.

What would the plot look like if the data were logarithmized?

```
plot(log(x),log(y))
```

You can see the result in Figure 6.11. Still not optimal for a linear regression, but you could make an attempt. That is exactly what we are trying to do now:

```
fit_log <- lm(log(y)~log(x))
```

The residuals approach a normal distribution, and the R-values of the model, while not exhilarating, are not unusable:

```
> summary(fit_log)
Call:
lm(formula = log(y) ~ log(x))
Residuals:
```

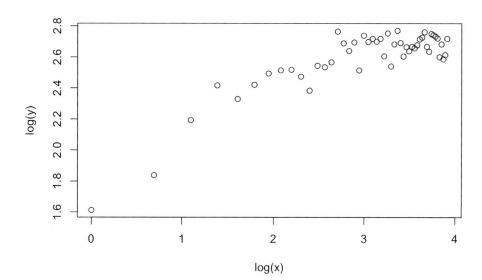

FIGURE 6.11
The point cloud with logarithmized data.

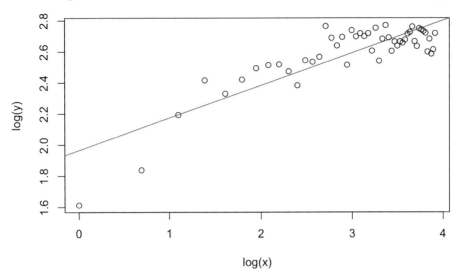

FIGURE 6.12
Plot of the regression line through the logarithmized data.

```
Min 1Q Median 3Q Max
-0.35155 -0.06342 0.00124 0.08639 0.23283
Coefficients:
             Estimate Std. Error t value Pr(>|t|)
(Intercept) 1.96428 0.05624 34.92 < 2e-16 ***
log(x) 0.20945 0.01816 11.53 1.93e-15 ***
---
Signif. codes: 0 '***' 0.001 '**' 0.01 '*' 0.05 '.' 0.
1 ' ' 1
Residual standard error: 0.113 on 48 degrees of freedom (
1 observation deleted due to missingness)
Multiple R-squared: 0.7349, Adjusted R-squared: 0.7293
F-statistic: 133 on 1 and 48 DF, p-value: 1.933e-15
```

The plot of the regression line also shows that the model is not optimal, but also not completely off. However, heteroscedasticity can also be observed here, see Figure 6.12.

I do not necessarily recommend this procedure; at least, you should check the described values (distribution of residuals, R-value) after the application to see if you really get a better result. It is also important that you always have to logarithmize new data points first and feed the result back again.

6.1.7 Logistic Regression

In the linear regression in our examples, we have a number for both the dependent variable and the independent variable. But, what if we don't have a number as the dependent variable, but a category?

The scale level

This is a good time to talk about scales, because scales determine what computational operations are possible with what data and when categories can be *translated* into numbers. For example, a variable that contains hair colors is nominally scaled. We can calculate frequencies here, but nothing more.

The next level up are ordinally scaled variables. These can be numeric, but the distances are not necessarily unique. An example is a sports league ranking: Is the team on the fourth place really four times worse than the team on the first place? Categorical data can also be ordinally scaled; high school graduation is an example. Clearly, a university degree is further away from a secondary school diploma than a high school diploma, but even if we replace the degree with years of education or schooling, the distances are not unambiguous.

The next level up are interval-scaled variables. An example is the temperature measured in Celsius: There is a zero point, but it is not an absolute zero point, because in Fahrenheit, the zero point is different.

The highest level is the ratio scaled variables. These have an absolute zero point. An example is age: no one is younger than 0 years.

The scale is not always unambiguous. Postal codes, for example, are clearly numerical, but the distance between the individual postal codes can be wonderfully argued about. Of course, the 2 as the first digit of the postal code is further away from the 8 than the 4.

Moreover, there may be more than one independent variable and, of course, more than one dependent variable. This is where analysis methods such as analysis of variance (also ANOVA for ANalysis Of VAriance) or multivariate analysis of variance (MANOVA for Multiple ANalysis Of VAriance) come into play.

6.2 Anomaly Detection

In the previous section, we looked at relatively simple forecasts. If more of one, then more of the other. Sometimes, however, we are also interested in deviations from the forecast. To do this, however, we must first identify what exactly is a deviation and what belongs to the usual patterns when we are not dealing with a simple regression curve.

6.2.1 Time Series Analyses

Let's say you have a website where users can communicate with your customer support via chat windows. Surely, it doesn't make much sense to split your support team into three shifts so that you can serve your customers 24 hours a day, unless you have built a global store. For that, it would be good to know how many visitors you should expect on your website, and when.

One way to visualize this is to plot the number of visitors per hour with time on the x-axis, and number on the y-axis. Many web analytics programs do just this, which then looks something like what is shown in Figure 6.13.

In this example, we also see that there are fewer users towards the weekend, and then on Monday morning, users come back to the site. This site has a business background, so visitors come primarily during office hours. How could we tell if the drop in visitors on the weekend is a pattern or an anomaly? First, let's keep in mind that we are looking at measures over a period of time, so the data is a sequence of measures ordered by time. This is called a time series.

We first turn to another data set that comes with R, USAccDeaths, which lists the monthly number of accidental deaths in the United States from 1973 to 1978.

```
> require(datasets)
> head(USAccDeaths)
[1] 9007 8106 8928 9137 10017 10826
```

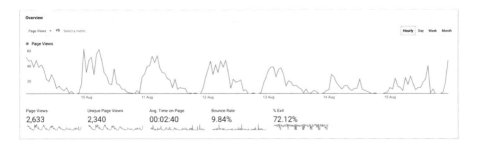

FIGURE 6.13
Screenshot from Google Analytics with a time series.

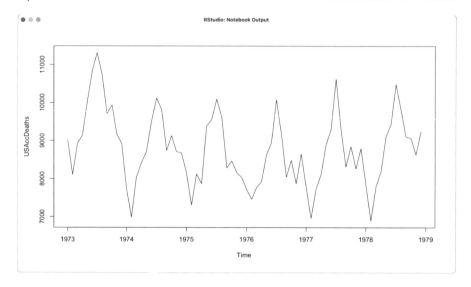

FIGURE 6.14
The "USAccDeaths" record.

This is not a real data frame, as we can find out with str():

```
> str(USAccDeaths)
 Time-Series [1:72] from 1973 to 1979: 9007 8106 8928 9137
     10017 ...
```

It is already a time series object! This allows us to plot it easily with plot() (otherwise we would have to use plot.ts()):

```
> plot(USAccDeaths)
```

The result can be seen in Figure 6.14 can be seen. We see the most frequent number in a month over the course of 1973; there always seems to be a peak in the middle of the year, and the low point is just after New Year's Day. We seem to see some kind of seasonality.

A time series can be decomposed into components, namely:

- the trend, i.e. how the data generally develops

- recurring fluctuations (seasonal)

- random fluctuations (random) that cannot be explained by the trend or the season

You can examine these components with the decompose() function:

```
decomposed <- decompose(USAccDeaths)
plot(decomposed)
```

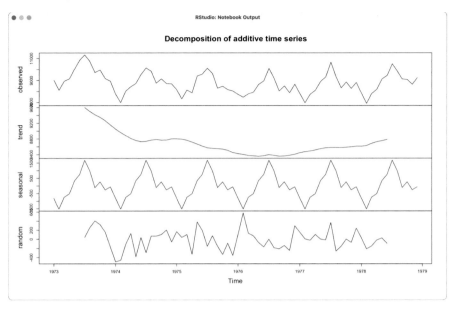

FIGURE 6.15
Decomposition of time series data into its components.

The plot of this division can be examined in Figure 6.15. Here, the original plot of the data is shown in the first row, the trend in the second, the seasonality in the third, and the noise in the fourth. Noise is defined here as meaningless data that has arisen by chance.

The seasonality is obviously uniform; the trend is first downward and then slightly upward again. The question now is how to convert these data and their components into a forecast.

6.2.2 Fitting with the Forecast Package

The package forecast contains the function ets(), which is used for Error, Trend, Seasonal. We start with a simple command (after you have installed the forecast package):

```
library(forecast)
fit <- ets(USAccDeaths)
plot(forecast(fit))
```

Ciao forecast, hello faible!

At the time of manuscript writing, the forecast package has

been "retired". However, bugs are still being fixed, and the documentation is being updated. The maintainer of the forecast package recommends the package faible, which, however, has a completely different philosophy. You can read more about this on the website for the book at https://alby.link/datascience.

The plot can be seen in Figure 6.16. Important: The ets() function creates a model (str() says this is a list); it does not yet create a forecast. This is only created with the command forecast().

This forecast consists not simply of a line that continues the previously observed data, but an interval around the line (forecast interval). To understand this in more detail, let's take a closer look at the ets() and forecast() functions.

First of all, we plot the model fit:

```
plot(fit)
```

and obtain the plot in Figure 6.17. It looks a bit like the plot in Figure 6.16 with the difference that we are talking about level and not trend, and the line in level seems to be much more granular. Also, we are missing what was called random in the other plot. There seems to be a difference in how time series data can be broken down (or decomposed) into individual components.

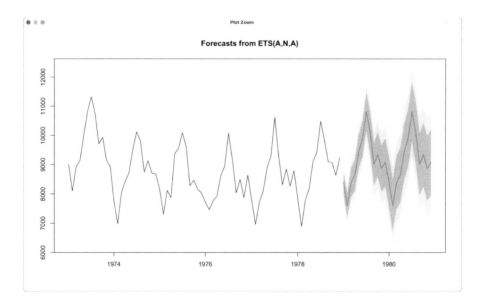

FIGURE 6.16
A forecast based on the basic settings of ets().

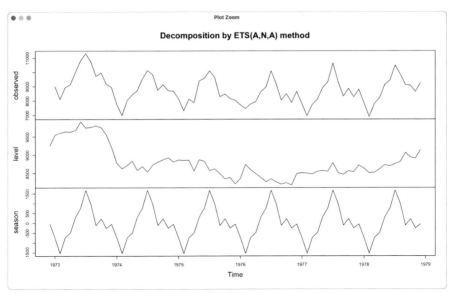

FIGURE 6.17
Decomposition with the ETS model.

decompose() uses a moving average, whereas ets() calculates a weighted average. These weights decrease exponentially with time, based on a "smoothing" parameter, so it is also called exponential smoothing.

But, that is not enough. If we output the summary for the model, we see more parameters:

```
> summary(fit)
ETS(A,N,A)
Call:
ets(y = USAccDeaths)
Smoothing parameters:
alpha = 0.5946
gamma = 0.002
Initial states:
l = 9248.3628
s = -51.3449 -255.3528 218.2901 -121.771 970.7387 1683.237
756.092 306.4212 -489.5627 -739.9004 -1537.792 -739.0552
sigma: 292.6907
AIC AICc BIC
1140.145 1148.716 1174.295
Training set error measures:
ME RMSE MAE MPE MAPE MASE ACF1
Training set -2.612086 262.698 202.6369 -0.08274519 2.32238
```

```
0.4634348 -0.007898621
>
```

We won't go through all the values (because that would also require me to tell you a bit more about state-space models), so here are the most important statements:

- First of all, we see the letters A, N, A after the call. The letter on the first position stands for the error type, the second letter for the trend type, and the third letter for the seasonality type. The following letters are possible:

  - N: none
  - A: additive; means that trend and seasonality add up to explain the development
  - M: multiplicative; means, very roughly, that trend and seasonality are multiplied by each other and thus provide for strong fluctuations
  - Z: automatically selected

- In the example, it is an additive error type, no trend type, and an additive seasonality type.

- alpha stands for the base value, and the higher alpha is, the more weight is given to the last observed values.

- beta does not occur here, but stands for the trend value: the higher beta is set, the more the trend data is taken into account.

- gamma stands for the seasonality component.

- AIC: Akaike's Information Criterion; see later.

- AICc: Corrected Akaike's Information Criterion; see later.

- BIC: Bayesian Information Criterion; see later.

The last values are parameters for comparing the model quality. AIC values are compared with AIC values of other models, so you do not compare an AIC value with a BIC value. So, for example, specify different values for alpha once:

```
fit <- ets(USAccDeaths, alpha = 0.5)
```

You will see that the predictions change slightly as you plot the model. Likewise, you can change the model in terms of the three letters:

```
fit <- ets(USAccDeaths, model = "MAA")
```

Again, the forecast changes slightly, as seen in the plot.

The model is one thing; the forecast is another. Here, too, you can set parameters. We will now use an example in the dplyr notation:

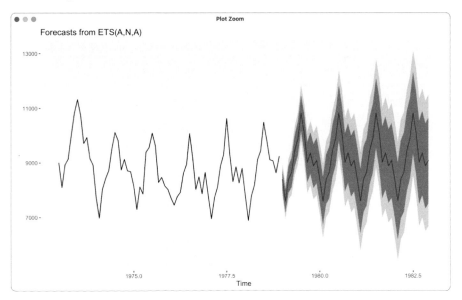

FIGURE 6.18
Autoplot of a forecast for 48 months.

```
USAccDeaths %>%
+ forecast(h=48) %>%
+ autoplot()
```

You can see the result in Figure 6.18. In this example, we have specified the forecast to be 48 months. Other frequently used parameters are:

- level: the confidence level

- robust: TRUE or FALSE; robustness against missing values or outliers

Models with ets() are obviously not a simple story, as you have to think about what your data and its decomposition look like, and formulate conclusions based on that. Do you think seasonality and trend are mutually reinforcing? Do the latest data have more relevance for you? In the simplest case, use the default parameters, but any additional knowledge can help you refine your model.

7

Clustering

How can patterns be identified in data from which groups can be derived, for example, customer segments? Clustering methods provide the answer. While in classification, we already know what spam looks like and can use existing spam to train our classifier, in clustering, we look for new patterns that can divide a data set into groups.

7.1 Hierarchical Clustering

Hierarchical clustering is an example of unsupervised machine learning. The algorithm is based on the calculation of the distances between all data points with respect to their characteristic values and the resulting groups, which are called clusters.

7.1.1 Introduction to the Algorithm

In the previous chapter on forecasting, we first talked about how distances help us to find the ideal straight line through our data points. Hierarchical clustering also works with distances, albeit in a slightly different way. While regression is about predicting what a value y will look like when x takes on a certain value, clustering is about finding patterns in the data that will lead us to group the data points in a meaningful way. I would like to illustrate this with a simple example.

Imagine the situation at a school: There are about 800 students who are between 10 and 20 years old and between 1.30 meters and 2 meters tall. In addition, there are about 100 teachers who are between 30 and 65 years old and between 1.60 meters and 2 meters tall. We already know that we actually have two groups, teachers on the one hand and students on the other – let's call them "teachers" and "school children," even though some of the "school children" may be young adults. How would an algorithm proceed that does not yet know the concept of the two groups? Using the age, we would have two clearly separated groups, but with the size, it is a bit more complicated. A pupil can be taller than a teacher, but not older.

To simplify the first step, we first focus on age, because in order to be able

DOI: 10.1201/9781003426363-7

to identify clusters, that is, which elements are so similar that they can be grouped together in a cluster, the elements must first be compared with each other. Since we have numerical values here, this is easy. A person who is 60 years old is further away from a person who is 15 years old than a person who is 40 years old. But since we cannot imagine 900 people and their distances from each other in our head, we simplify the example even further and take a very small school with 20 school children and 5 teachers, and again, we focus on age.

First, we generate random numbers for the ages of both groups:

```
> ages_pupils <- sample(10:20, 20, replace = TRUE)
> ages_teachers <- sample(30:65, 5, replace = TRUE)
> ages <- c(ages_pupils, ages_teachers)
> ages
 [1] 16 10 18 17 19 17 17 11 20 15 12 11 18 19 10 12 10 12 19
 18 54 65 52 54 37
>
```

We see that the first schoolchild (age 14) has a distance of 3 years from the second schoolchild (age 11), likewise from the third, and then 4 from the next schoolchild (age 18). The distance is simply determined by subtracting the second value from the first. Let's look at the data in a histogram:

```
hist(ages, breaks=20)
```

You can see the histogram in Figure 7.1. Clearly, we have here two groups separated by the age intervals.

In the next step, we would have to calculate the distance of each school child and each teacher to each other. Nicely enough, there is already a function that allows us to calculate the distances between each person automatically, so we don't have to do it by hand. The result is called a *distance matrix*, sometimes also *dissimilarity matrix*, here shortened to fit on the page:

```
> ages.dist <- dist(ages)
> ages.dist
    1  2  3  4  5  6  7  8  9 10 11 12 13 14 15 16 17 18 19
2   6
3   2  8
4   1  7  1
5   3  9  1  2
6   1  7  1  0  2
7   1  7  1  0  2  0
8   5  1  7  6  8  6  6
9   4 10  2  3  1  3  3  9
10  1  5  3  2  4  2  2  4  5
11  4  2  6  5  7  5  5  1  8  3
```

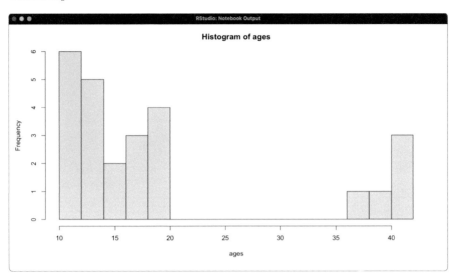

FIGURE 7.1
Histogram of age data.

```
12  5  1  7  6  8  6  6  0  9  4  1
13  2  8  0  1  1  1  1  7  2  3  6  7
14  3  9  1  2  0  2  2  8  1  4  7  8  1
15  6  0  8  7  9  7  7  1 10  5  2  1  8  9
16  4  2  6  5  7  5  5  1  8  3  0  1  6  7  2
17  6  0  8  7  9  7  7  1 10  5  2  1  8  9  0  2
18  4  2  6  5  7  5  5  1  8  3  0  1  6  7  2  0  2
19  3  9  1  2  0  2  2  8  1  4  7  8  1  0  9  7  9  7
```

We can reproduce our observation for the first values here as well. We also see in the last rows that the distance between the teachers and the schoolchildren is larger, compared with the distances between students in their group and teachers in theirs. We plot a first *dendrogram* (see Figure 7.2) that visualizes the distances between the clusters.

```
ages.hc <- hclust(ages.dist, method = "complete")
plot(ages.hc)
```

The *dendrogram* in Figure 7.2, we will cover the concept in a few paragrahs, clearly shows two clusters, teachers on the left and school children on the right. The numbers represent the position in the list we generated earlier. The teachers among themselves as well as the school children among themselves are each further subdivided.

Let's take a closer look at that as well. On the left in the teacher cluster, we see the following subclusters:

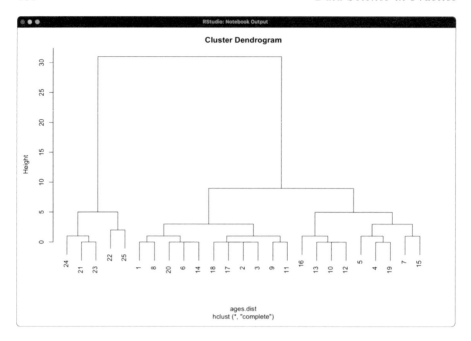

FIGURE 7.2
Our first dendrogram.

- a subcluster with two further subclusters:

  - Data point 24: 41 years old
  - Data points 21 and 23: both 42 years old

- a subcluster with data points 22 and 25, 37 and 39 years old

Data points 21 and 23 have no distance between them because the individuals are the same age, we put them into one cluster. Data point 24 has a distance of 1 from data points 21 and 23, so it also makes sense to place it close to points 21 and 23. The teachers aged 37 and 39 have a distance of 2 from each other and a distance of 2 from data point 24, but since data point 24 is closer to data points 21 and 23, the two teachers are further away from it.

A dendrogram is calculated from the bottom up. We first look at the individual data points and their distances from each other, form the first clusters from them, and then, based on the distances of these clusters from each other, the next higher clusters are formed until we have only one cluster at the very top. This bottom-up procedure is also called the *agglomerative method*. There are also top-down methodsdivisive), in which the approach is top-down or *divisive*.

This already explains all the essentials of hierarchical clustering. But, how do we calculate distances when we have more than one variable?

7.1.2 The Euclidean Distance and its Competitors

In our first clustering attempt, we used only one part of the data set, age. Here, the distance is quite simple to calculate; the distance of the persons to each other in relation to their age. However, we have a second variable, size. For the teachers, we again generate random numbers, and we use centimeters as the unit:

```
> heights_teachers <- sample(160:200, 5, replace = TRUE)
> heights_teachers
[1] 186 164 164 197 178
```

From this we create a data frame:

```
> teachers <- data.frame(ages_teachers,heights_teachers)
```

With school children, it's not quite so simple with height. Age and height correlate, which means an 11-year-old with a height of 2 meters will occur rather rarely, as will an 18-year-old with a height of 1.40 meters. Again, random numbers are generated that we will adjust manually:

```
> heights_pupils <- c(175,147,144,187,180,156,166,164,159,187,
  153,182,177,158, 160,186,138,134,168,150)
> pupils <- data.frame(ages_pupils,heights_pupils)
```

Now, the two data frames are combined to build the distance matrix. Before, we need to clean the data set as the column names are different. For rbind(), both data frames must have the same column names. Again, the output is shortened here:

```
pupils <- pupils %>%
rename(age = ages_pupils, height = heights_pupils)
teachers <- teachers %>%
rename(age = ages_teachers, height = heights_teachers)
> school <- rbind(pupils, teachers)
> dist(school)
        1 2 3 4 5 6 7 8
2 28.160256
3 31.144823 3.000000
4 12.649111 40.607881 43.566042
5 5.830952 33.541020 36.496575 7.071068
6 19.026298 9.219544 12.165525 31.400637 24.331050
7 9.055385 19.416488 22.360680 21.213203 14.142136 10.198039
```

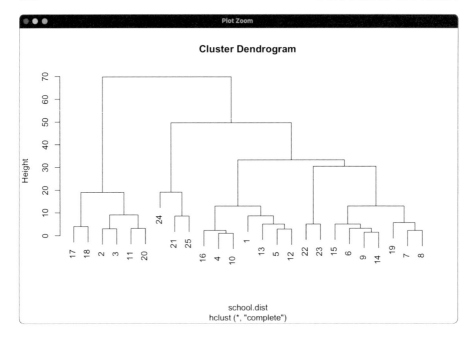

FIGURE 7.3
The dendrogram with age and size, but the data has not been scaled before.

---

rbind() and cbind() from the base R

We could have used the join command from the Tidyverse here, but I also wanted to introduce you to an alternative from Base R. rbind() stands for row bind; there is also cbind() for column bind. Sometimes, you can work a little bit faster with these commands.

After that, we use hclust() again and plot a dendrogram, which you can see in Figure 7.3. Three teachers are still hanging together, but are then connected to schoolchildren in a cluster. Likewise, two other teachers are connected to school children.

I have only reproduced part of the distance matrix here, but we can already see that the distances look somewhat different than before. How do these values come about?

Previously, for one variable, we had simply subtracted one value from the second value. Now, we have two variables, and in fact, we don't do things that differently. We can now simply subtract the data points from each other as before and add the result.

Let's go through this with the example of the first two school children. Child 1 is 14 years old and 175 cm tall. Child 2 is 11 years old and 147 cm tall. As we saw before, the distance is 3. But – and normally someone would now scream out in class – sometimes, when we subtract values, a negative number should have come out in the distance matrix, for example at data points 1 and 4. As you have already seen in Chapter 5, "Exploratory Data Analysis", there is a little trick to get rid of the sign, and that is by squaring the result of the subtraction first and taking the root later. As a formula, write it like this for two variables:

$$\sqrt{(a-b)^2 + (c-d)^2}$$

Here, a and b stand for the data points 1 and 2 of the first variable and c and d for the data points 1 and 2 of the second variable. Let's insert the values from our example:

$$\sqrt{(14-11)^2 + (175-147)^2} = \sqrt{9+784} = \sqrt{793} = 28.16$$

We look again in the distance matrix calculated earlier and see that we get the same result. If we had three variables, then we could add another term, and if we had a fourth variable, we could add another term, and so on.

What we have calculated here is called *Euclidean distance*. It is, roughly speaking, the "flight line" between two data points, if we think of it like a map in a coordinate system with two axes. The Euclidean distance is the default when you use the dist() function. Besides Euclidean distance, there are other options, namely *maximum, manhattan, canberra, binary,* and *minkowski* (to be specified with method = OPTION). I won't explain all options in detail, but you can certainly imagine some of them yourself. The Manhattan distance, as an example, goes "around the block", so to speak, so it does not have the shortest connection via the flight line. As a small rule of thumb, you can remember that for most cases you can use Euclidean distance. If you have many dimensions, Manhattan distance is advised. This has to do with the curse of dimensionality [1]. The Minkowski distance is a generalized form of the Euclidean distance and the Manhattan distance. You can read more about it on my blog. For starters, the Euclidean and Manhattan distances are enough for you.

Perhaps you have noticed another unattractive feature: The age ranges somewhere between 10 and 43 years, but the height ranges between 130 and 200 centimeters. In other words, height may have a stronger influence on the distance calculation. We still want to correct this.

7.1.3 The Distance Matrix, but Scaled

As just noted, we have used dist() without taking into account the differences in the data, which can lead to bias if the variables are different in their proportions. We can use the scale() function here, which calculates the z values and thus standardizes the data points. So, it subtracts the arithmetic mean of the column from each value in a column and then divides the result by the standard deviation of the column. You could also say that the standardized value indicates by how many standard deviations a value deviates from the mean. Example:

```
> mean(school$age)
[1] 19.88
> sd(school$age)
[1] 10.78317
```

Then, the standardized age value of the first school child:

```
> (school[1,1] - mean(school$age)) / sd(school$age)
[1] -0.5452944
```

We are checking this:

```
> head(scale(school))
            age height
[1,] -0.5452944 0.5116691
[2,] -0.8235058 -1.1542304
[3,] -0.8235058 -1.3327196
[4,] -0.1743458 1.2256261
[5,] -0.2670830 0.8091512
[6,] -0.6380315 -0.6187627
```

The values match (see the first age value), so scale() does a good job, as you would expect. What effect does this have on our clustering?

```
> school.dist <- dist(scale(school))
> school.hc <- hclust(school.dist)
> plot(school.hc)
```

The dendrogram in Figure 7.4 shows that the teachers and the school children are now more clearly separated again, but we now have three clusters depending on where we cut. Some school children are closer to teachers than to other school children. This also makes sense, because some school children are taller than others.

7.1.4   The Dendrogram

We have now seen several dendrograms, but there is more to discover here. A dendrogram is a visualization of a distance matrix, so to speak, and can be read as showing the distances between clusters on the y-axis; the distances on the x-axis have nothing to say. If we take a closer look at the dendrogram in Figure 7.4 more closely, however, we cannot necessarily follow this explanation: Data

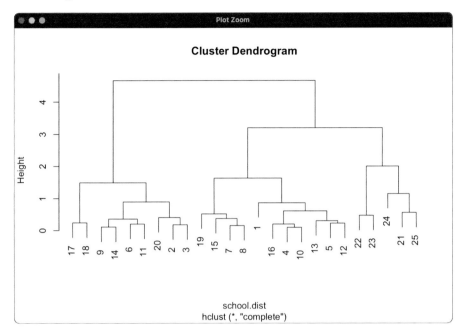

FIGURE 7.4

The dendrogram after standardization of the data.

points 21 and 23 have a distance of 0 from each other, and so their meeting on the y-axis is plotted at the zero level. This point is also called a *node*. Data points 22 and 25 have a distance of 2 from each other, and their node is plotted on the y-axis at the height of 2. But, how is the distance between these two nodes calculated? The distance of the data point 22 to the points 21 and 23 is 5, whereas the distance of the data point 25 to the two points is only 3.

When creating the dendrogram, a parameter was specified, complete, which stands for complete linkage.and this parameter is used to select the maximum distance, i.e. 5. Therefore, the node for the two clusters is drawn at 5. In the dendrogram, information is lost that we can only see in the distance matrix, because the distances between data point 24 and points 22 and 25 from the distance matrix do not make sense. If the data were scaled, then the original distances cannot be read here.

We have the option to display the clusters separated by color, but to do this, we must choose how many clusters we want to divide the data into. This is done with the function rect.hclust():

```
school.hc <- hclust(school.dist, method = "complete")
plot(school.hc)
rect.hclust(school.hc, k=3, border="red")
```

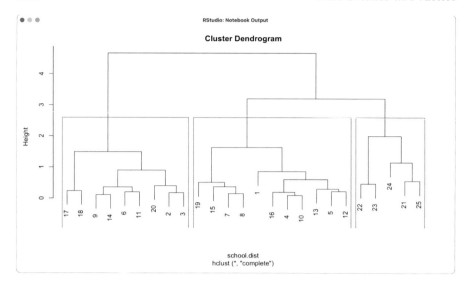

FIGURE 7.5
Clusters drawn in the dendrogram.

The result can be seen in Figure 7.5. In this case, I chose three clusters. How many clusters make sense depends on the data, but also on the application of the results (see Section 7.1.6).

As with abline() in the previous chapter, rect.hclust() must be called in a notebook immediately after plot().

7.1.5 Dummy Variables: What If We Have No Numerical Data?

This is an interesting problem that we will also encounter in the other chapters. In the school example, for example, one could insert the highest school degree as a variable. We could even talk about a kind of distance, because a university degree is further away from an elementary school degree than a high school diploma. We could assign numbers – a 1 for an elementary school degree, a 2 for a high school degree, and so on. But, what if we want to include gender, for example, and have categories like male/female/diverse? Diverse is no further away from male than it is from the female category. This approach does not work.

One way to solve this problem is to use *dummy variables*. Instead of assigning a number for each expression, which, as we have just seen, is not always possible, we create a variable for each expression and simply say whether it is present or not. A 1 means that the expression is present, a 0 means that it is not present. An example for "male":

male	female	diverse
1	0	0

However, one column results from the other two columns, because if someone is male and not female, then this person cannot be diverse. However, if "male" and "female" were both 0, then the person would automatically be diverse. We always need one dummy variable fewer than we have characteristic values. There are also functions for creating dummy variables, which you will learn about in the chapter on classification.

Another possibility is offered by the daisy() function. It belongs to the package cluster and can also calculate distances for non-numeric data. For this, the Gower distance is used, which looks at text variables to see if there are equal elements, and then assigns a 0, and a 1 if two data points are equal. Be careful here as well!

7.1.6   What Do You Do with the Results?

This question arises with every algorithm, but with clustering in particular, since the respective algorithm is not trained, and we do not necessarily know what may come out. The results do not always make sense, and if we stick to our example, this is understandable. The example is also a bit far-fetched, because we already know the differences. There are school children and teachers, and they are in different age groups. But it also clearly shows how important it is to think in advance about which variables make sense to include. We have already discussed the importance of action relevance, sometimes called "actionability" in new German, in previous chapters, but here it is even more important. Suppose you now discover 20 clusters in the customer data in your Customer Relationship Management system (CRM), do you have the ability to run 20 different campaigns?

As mentioned earlier, it is not always possible to do something with the results. Please don't be disappointed if the result is not meaningful. Sometimes, there is no result because there is no pattern in the data, even if you have invested a lot of brainpower in selecting the variables. That, too, is part of the job and a realization – albeit one that your clientele, both internal and external, won't like to hear. Especially not when a lot of time and money has been invested in an analysis.

7.2   k-Means

k-Means is a popular algorithm in every Data Science tutorial and probably one of the best known clustering algorithms. The name already says what it is about: k stands for the number of clusters, and Mean, as we already know, stands for an average.

7.2.1   How the Algorithm Works

First of all, k center points (*centroids*) are placed in the data points. This is done randomly, so for the same data set and multiple runs of k-Means, different results may come out. Now, the actual center points within each cluster are computed by calculating the distances between the data points and the center point. This reassigns the data points to the clusters, so the midpoints "wander" through the data points, so that a data point can also be assigned to a different cluster. The process is repeated again and again until there are no more changes.

We again use the previous data set with school children and teachers and tell k-Means to identify three clusters; I will augment R's output with my comments:

```
> school.fit <- kmeans(school.scaled, 3)
> school.fit
K-means clustering with 3 clusters of sizes 11, 5, 9
```

The given number of clusters is achieved with different cluster sizes. This is exciting, because we usually want to have clusters of reasonably equal size, rather than two large ones and one very small one. Now comes the data of the centers:

```
Cluster means:
age height
1 -0.6886154 -0.9054272
2 1.8844188 0.6782591
3 -0.2052582 0.7298227
```

Finally, we obtain the assignment of each data point to a cluster:

```
Clustering vector:
[1] 3 1 1 3 3 1 3 1 1 3 1 3 1 1 3 1 1 3 1 2 2 2 2 2
```

I will discuss the following lines below:

```
Within cluster sum of squares by cluster:
[1] 3.453850 3.053011 2.030958
(between_SS / total_SS = 82.2 %)
```

Finally we get a hint which data we can use in the created object.

```
Available components: [
1] "cluster" "centers" "totss" "withinss" "tot.withinss"
```

```
"betweenss" "size" [
8] "iter" "ifault"
>
```

We plot the data of the data set once along with the cluster centers:

```
school.comp <- as.data.frame(cbind(school.scaled,
    school.fit$cluster))
school.comp <- school.comp %>%
    rename(Cluster = V3)
    ggplot() +
    geom_point(data = as.data.frame(school.comp),
        aes(age, height, shape = factor(Cluster)), size=3) +
    geom_point(data = as.data.frame(school.fit$centers),
        aes(age, height), color = "red", shape="diamond", size=4)
```

You can see the result in Figure 7.6. The data points for the school children have been divided into two clusters, correctly one with smaller and younger school children and one with larger and older school children. Added to this is the cluster for the teachers on the far right.

Now, let's use two clusters; for this we only have to replace the 3 by a 2 in the code, so that a plot like in Figure 7.7 should be obtained.

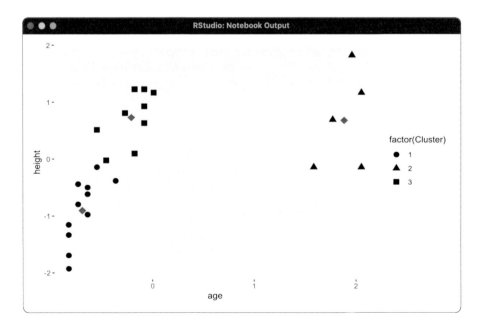

FIGURE 7.6
Plot of the data points with their centroids.

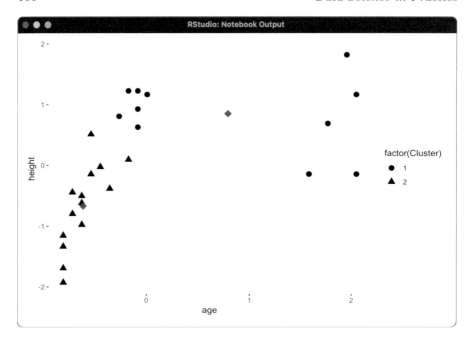

FIGURE 7.7
kmeans() with two cluster.

Interestingly, k-Means does not "recognize" the teachers as a homogeneous cluster, but includes the older and larger school children. However, we can also give k-Means initial data for the centroids, which we estimate with respect to the data, for example like this:

```
start <- matrix(c(2, 1, -0.5, 0), 2)
school.fit <- kmeans(school.scaled, centers = start)
```

In this case, k-Means remains in the clusters we would have expected as well, cf. Figure 7.8.

7.2.2   How Do We Know k?

One way to find a suitable k may be to first perform the hierarchical clustering. In this example, we use a data set on wines that you can download from https://alby.link/wine [56]. It contains the results of a chemical analysis and a variable that says what type of wine it is. In total, there are three groups of wines.

First, we try hierarchical clustering with this data set and see if we can identify an ideal number of clusters:

```
> wine.dist <- dist(scale(wine[,-1]))
```

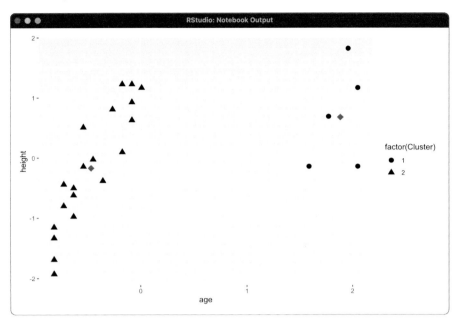

FIGURE 7.8
The clusters with centroids oriented to predefined starting points.

```
> wine.hc <- hclust(wine.dist)
> plot(wine.hc)
> rect.hclust(wine.hc, k=3, border="red")
```

▲	X1	X2	X3	X4	X5	X6	X7	X8	X9	X10	X11	X12	X13	X14
1	1	14.23	1.71	2.43	15.6	127	2.80	3.06	0.28	2.29	5.64	1.040	3.92	1065
2	1	13.20	1.78	2.14	11.2	100	2.65	2.76	0.26	1.28	4.38	1.050	3.40	1050
3	1	13.16	2.36	2.67	18.6	101	2.80	3.24	0.30	2.81	5.68	1.030	3.17	1185
4	1	14.37	1.95	2.50	16.8	113	3.85	3.49	0.24	2.18	7.80	0.860	3.45	1480
5	1	13.24	2.59	2.87	21.0	118	2.80	2.69	0.39	1.82	4.32	1.040	2.93	735
6	1	14.20	1.76	2.45	15.2	112	3.27	3.39	0.34	1.97	6.75	1.050	2.85	1450
7	1	14.39	1.87	2.45	14.6	96	2.50	2.52	0.30	1.98	5.25	1.020	3.58	1290
8	1	14.06	2.15	2.61	17.6	121	2.60	2.51	0.31	1.25	5.05	1.060	3.58	1295
9	1	14.83	1.64	2.17	14.0	97	2.80	2.98	0.29	1.98	5.20	1.080	2.85	1045
10	1	13.86	1.35	2.27	16.0	98	2.98	3.15	0.22	1.85	7.22	1.010	3.55	1045
11	1	14.10	2.16	2.30	18.0	105	2.95	3.32	0.22	2.38	5.75	1.250	3.17	1510
12	1	14.12	1.48	2.32	16.8	95	2.20	2.43	0.26	1.57	5.00	1.170	2.82	1280
13	1	13.75	1.73	2.41	16.0	89	2.60	2.76	0.29	1.81	5.60	1.150	2.90	1320
14	1	14.75	1.73	2.39	11.4	91	3.10	3.69	0.43	2.81	5.40	1.250	2.73	1150
15	1	14.38	1.87	2.38	12.0	102	3.30	3.64	0.29	2.96	7.50	1.200	3.00	1547

FIGURE 7.9
The wine dataset.

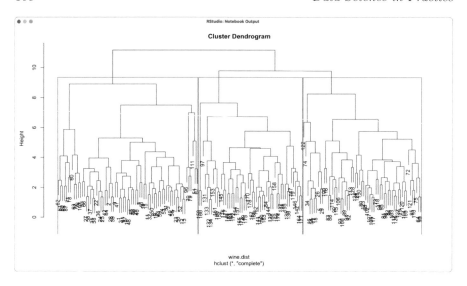

FIGURE 7.10
The dendrogram with the three clusters of wines.

The result in Figure 7.10 shows us three or four clusters, depending on where we cut. In the diagram, we cut at 3 (knowing that we have three clusters).

Another and more commonly used way to find out the number of clusters is in the so-called *scree test*, also called the *elbow test*. We first create a data frame with scaled values:

```
wine.scaled <- as.data.frame(scale(wine[,-1]))
```

And now, we use k-Means to calculate the variability within the clusters:

```
wss <- (nrow(wine.scaled)-1)*sum(apply(wine.scaled,2,var))
for (i in 2:10) wss[i] <- sum(kmeans(wine.scaled,
centers=i)$withinss)
plot(1:10, wss, type="b", xlab="Number of Clusters",
ylab="Within groups sum of squares")
```

The result is a plot like that in Figure 7.11. You can see that the variability has a bend starting at 3. In other words, unlike hierarchical clustering, it is clear that we should take three clusters. It is also clear from the plot what this visualization has to do with an elbow. The advantage of this approach is obvious: by already using k-Means to calculate variability, we have a related determination of the number of clusters to the algorithm.

Now, let's run k-Means "properly":

```
> k.means.fit <- kmeans(wine.scaled, 3) >
```

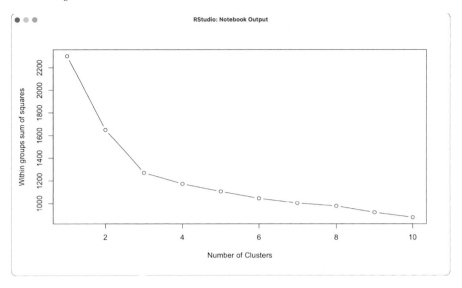

FIGURE 7.11
The result of the elbow or scree test.

```
table(wine$X1,k.means.fit$cluster)

     1 2 3
1 0 59 0
2 65 3 3 3 0 0
48
```

We see that we have correctly classified almost all wines. At first sight, the sorting looks a bit strange, because after all, we have sorted 59 wines from group 1 into group 2. But, k-Means does not know which group it was before, and at least sorts the wines into one group and does not distribute them to different ones. It is left to chance which number k-Means uses for each identified group. With the wines of group 2, on the other hand, we already see a few errors.

7.2.3 Interpretation of the Results

Unlike a dendrogram, we have no easy way to visualize the results here. While the dendrogram connects each data point somewhere and marks the distance on the y-axis, with k-Means, we may have a variety of dimensions that are not so easy to map.

But, there is also an alternative: In the package cluster there is the possibility to perform a cluster plot, in which the first two principal components are visualized. In Principal Component Analysis, several variables are

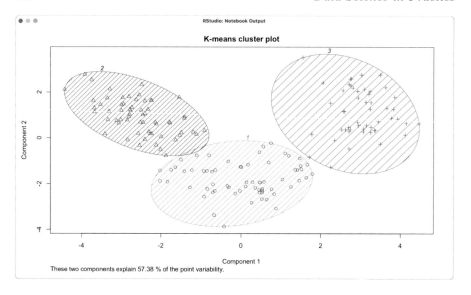

FIGURE 7.12
Cluster plot with two principal components.

clustered into a new variable. To achieve this, we look at which variables measure something similar or correlate. Principal component analysis is an interesting area that we unfortunately cannot give more attention to here; I highly recommend [15]. The cluster plot uses PCA to plot clusters in terms of the first two components. It is simply created using clusplot():

```
clusplot(wine, k.means.fit$cluster, color=TRUE, shade=TRUE,
labels=4, lines=0, main="k-means cluster plot")
```

The result can be seen in Figure 7.12 can be seen. We see our wines in three clusters. A little bit of one cluster protrudes into another, but none of the data points are in this intersection. The clusters seem to be very separable.

Nevertheless, we have wines that were ranked differently than expected. How does this happen? For one thing, data points can be very close to another cluster and thus be assigned to the wrong cluster. You can see this, for example, in Figure 7.13: Data point 62 is close to the cluster it actually belongs to.

On the other hand, it could be that a wine belongs to one group purely technically, i.e. in terms of its data, but is classified differently by wine connoisseurs. If you drink a Sauvignon Blanc from Marlborough in New Zealand and then one from California, you will probably taste big differences, yet both wines are Sauvignon Blancs.

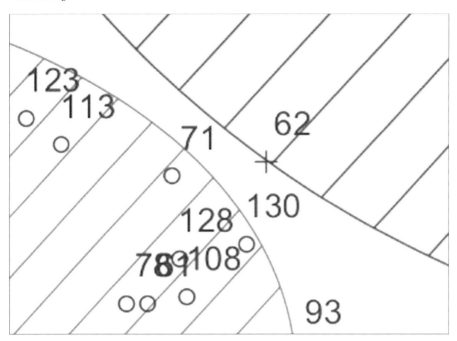

FIGURE 7.13
Data point 62 is in the wrong cluster.

7.2.4 Is k-Means Always the Answer?

Just because k-Means worked better than hierarchical clustering on the wine data set, it does not mean that k Means is always better. For example, we had slight problems with the school data set at first; only after we specified initial values for the centroids did it come out as we would have expected.

k-Means also has difficulties when the algorithm encounters dummy variables. Imagine that we have mostly 1s and 0s in a two-dimensional space, as in our school example. How will the midpoints align then? Hierarchical clustering may work better here, but it doesn't have to. However, the examples also show the importance of understanding how the algorithms work and how they will work with the data at hand.

8

Classification

Spam or not spam? This question is answered for us by algorithms of our mail providers and programs. How to solve this and other classification problems will be the subject of this chapter.

A spam detector is famouos example for a classifier. It is special in that we have only dealt with numeric data in clustering so far, but here we have to work with words. We will look at exactly how this works in Section 8.7, "Classification of Text." And, a second point stands out: If a spam mail is misclassified and lands in our mailbox, we get annoyed that the spam program didn't recognize it. But if we don't get an important mail because the spam program classified it as spam, then we get even more annoyed. As you have already learned, it depends on the context which error is worse. In this case, we might rather have a few spam mails in our inbox than miss an important mail. Of course, you can argue that you can just check the spam folder more often, but the expectation of many users is certainly different.

8.1 Use cases for classification

We also experience another example of classification every day, and that is when it comes to payments, whether in a store or even online, because as soon as we use a credit card or something similar, a system jumps in to see if the transaction is suspicious. Is something being purchased that is not typical for the credit card holder? Is it bought at a place where the card holder can't actually be, because he paid for his groceries at the supermarket just before? Good data sets usually exist here, as credit card holders are quick to report when they have not made a transaction. Again, the question is what is worse: seeing a problem with a payment where there is none, or not seeing a problem that clears the account. The first mistake is embarrassing, because the seller might think the account is not funded, but you might as well not care. In the worst case, you just stand at the checkout and can not pay. The other case, on the other hand, can be much worse, for example, if the account has been cleared and the credit card company initially refuses to refund the amount. Fraud systems can be purely rule-based, for example, in that the algorithm only kicks in when a certain amount is reached and then sounds the alarm

DOI: 10.1201/9781003426363-8

for certain store categories where problems have occurred more frequently. Fraudsters also learn and would then use other vulnerabilities. Therefore, it is important here to find other variables that could have an influence on a classification.

Equally well-known is the example of creditworthiness. Many transactions involve checking whether we are creditworthy, for example when it comes to buying an item on account (which is also already a type of credit) or even arranging payment by installments. Historical data can be used here as well, by including in the training of an algorithm both transactions that have been successful and those that have been problematic. When we register somewhere, a credit agency such as Schufa is usually asked how our creditworthiness is assessed.

Which is worse: if a loan is not granted and the customer may not buy the goods, or if a loan is granted but not repaid? That certainly depends on how strictly the algorithm acts. Many algorithms output a probability so that a threshold value can be determined above which no more credit is granted. This threshold value should then represent the best compromise between conducting business and preventing credit default.

Another example of classification is an algorithm designed to detect insurance fraud. Not all claims are legitimate. However, in some claims, damage is only faked, or the course of events is presented differently than it actually occurred, so that the insurance company should not have to pay at all. This costs the insurance industry a fortune, and in the end, everyone pays for it, because the loss ratio is priced in. Ideally, therefore, the claims application is already automated to see whether there are any suspicious facts that require further examination. To do this, previous claims are included in the training where it was determined that insurance fraud was involved.

This is where the problem of class imbalance, which I explained in Chapter 5, "Exploratory Data Analysis," arises. For our previous example, spam emails, we can assume that, unfortunately, there are enough spam emails to train with. And since we also receive enough legitimate emails, we should have no problem for this class either. Insurance fraud, on the other hand, is fortunately different, because most people are honest, so only a fraction of claims are problematic. However, you also don't realize in every case that someone is not honest, and in other cases you have a suspicion, but it can't be proven. In such a case with a class imbalance, models often just perform poorly, even if an attempt is made to "fill up" the smaller class with appropriate means.

Other possible applications:

- In a customer database, an algorithm is used to check which customers are more likely to cancel based on past cancellations by previous customers. What are the patterns of customers who have cancelled? For example, was there a cluster of complaints? These customers could be approached and a non-binding conversation about satisfaction held first, before possibly making a special offer.

- Similarly, an algorithm can be used in a customer database to classify whether some customers are more likely to be considered for a new product or service than other customers.

- Images of moles are used to classify whether a melanoma is present or not. This is a critical example insofar as here, a type II error (there is something there that you can't see) is extremely problematic. The company Fotofinder, for example, already offers systems to help dermatologists assess moles. What is exciting about this type of classification is that it is not initially a matter of numbers, but of image material that is converted into numbers.

Each of these use cases requires that first of all, an A/B test should be carried out to check to check whether the trained model delivers meaningful results outside the laboratory. However, this can also be problematic, because if there are already very few cases in a class, there will be even fewer cases if all classification objects are split.

8.2 Create Training and Test Data

We will use the same data set for all algorithms in this chapter. Therefore, it makes sense that we take a closer look at the data set and the division into test and training data. Depending on the algorithm, there may be some adjustments, but the basic concept is conveyed in this section.

Additionally, there is an example at the end of the chapter with a text data set; there, we will actually take a look at how a spam classifier works.

8.2.1 The Titanic Data Set: A Brief EDA

In 1912, the collision of the Titanic with the iceberg cost the lives of 1,514 of the 2,200 people on board. As macabre as it may sound, this data is often used for classification tutorials. You can find the Titanic dataset with the data of all passengers in different versions on the net; even in R there is already a data set for the Titanic passenger list. I use a prepared version of this data set in my courses, which you can download from https://alby.link/titanic. Once the data is downloaded (I recommend saving it in a directory like data), you can read it in using the readr library in the Tidyverse package as follows:

```
library(tidyverse)
titanic <- read_delim("data/titanic.csv", delim = ";",
escape_double = FALSE, trim_ws = TRUE)
```

The data contains only the passengers, not the crew of the Titanic; therefore, it has only 1,310 rows. In the data set, there is a column "survived"; here

pclass	survived	name	sex	age	sibsp	parch	ticket	fare
1	1	Allen, Miss. Elisabeth Walton	female	29	0	0	24160	211,3375
1	1	Allison, Master. Hudson Trevor	male	0,9167	1	2	113781	151,5500
1	0	Allison, Miss. Helen Loraine	female	2	1	2	113781	151,5500
1	0	Allison, Mr. Hudson Joshua Creighton	male	30	1	2	113781	151,5500
1	0	Allison, Mrs. Hudson J C (Bessie Waldo Daniels)	female	25	1	2	113781	151,5500
1	1	Anderson, Mr. Harry	male	48	0	0	19952	26,5500
1	1	Andrews, Miss. Kornelia Theodosia	female	63	1	0	13502	77,9583
1	0	Andrews, Mr. Thomas Jr	male	39	0	0	112050	0,0000
1	1	Appleton, Mrs. Edward Dale (Charlotte Lamson)	female	53	2	0	11769	51,4792
1	0	Artagaveytia, Mr. Ramon	male	71	0	0	PC 17609	49,5042
1	0	Astor, Col. John Jacob	male	47	1	0	PC 17757	227,5250
1	1	Astor, Mrs. John Jacob (Madeleine Talmadge Force)	female	18	1	0	PC 17757	227,5250

FIGURE 8.1
Extract of the columns of the Titanic dataset.

a 1 means the person survived and a 0 means the person did not survive. The goal for this chapter is to select from the available features those that allow us to predict whether or not a person survived (see Figure 8.1).

To explain the columns:

- "pclass": passenger class

- "survived": survived or not

- "name": name

- "sex": gender

- "age": age

- "sibsp": number of siblings/spouses on board.

- "parch": number of parents/children on board

- "ticket": the number of the ticket

- "fare": the price of the ticket

- "cabin": the cabin number

- "embarked": abbreviation for the port where the passenger embarked.

- "boat": number of the lifeboat

- "body": number of the body

- "home.dest": home town

We will now look at some of the features, first of all the distribution of survivors and non-survivors in terms of gender:

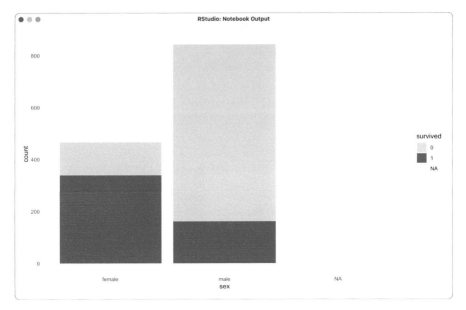

FIGURE 8.2
Stacked bar plot for gender and survival.

```
titanic %>%
    mutate(survived = as.factor(survived)) %>%
    ggplot(., aes(x=sex, fill=survived)) +
        geom_bar() +
        scale_fill_brewer(palette="Paired")+
        theme_minimal()
```

Obviously, while many more men were on the Titanic, at the same time, they died in excess of this ratio, as you can see in Figure 8.2. "Women and children first" is the saying on ships, let's see if this is also visible in the titanic data.

In this case, we take a boxplot; with the parameter varwidth=TRUE we cause the number of cases to have an influence on the width of the box by drawing it with a width proportional to the square root of the sum of the observations. The age still has the data type character, and here, we exchange the comma for a point before we convert the variable into a number.

```
titanic %>%
    mutate(survived = as.factor(survived)) %>%
    mutate(age = as.double(str_replace(age,",","."))) %>%
    filter(!is.na(age)) %>%
    ggplot(., aes(x=survived, y=age, fill=sex)) +
```

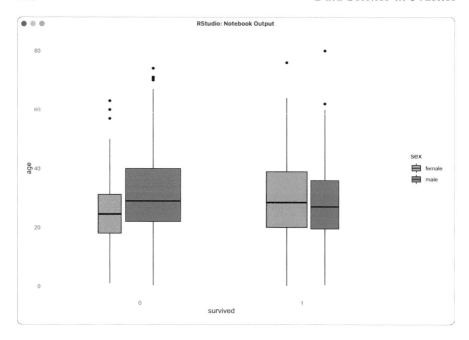

FIGURE 8.3
Boxplot with the variables gender and age.

```
geom_boxplot(varwidth=TRUE) +
theme_minimal()
```

Age also seems to have an influence; at least, slightly younger women seem to be more likely not to have survived the accident, and there is a small reverse trend among men. It is nice to see in this boxplot (Figure 8.3 that the varwidth parameter also differentiates the ratios of the sexes to each other.

Next, we add the passenger class:

```
titanic %>%
    mutate(survived = as.factor(survived)) %>%
    mutate(age = as.double(str_replace(age,",","."))) %>%
    filter(!is.na(age)) %>%
    ggplot(., aes(x=survived, y=age, fill=sex)) +
        geom_boxplot(varwidth=TRUE) +
        facet_grid(cols = vars(pclass))
```

We can clearly see that these three variables – gender, age, and passenger class – have an impact on survival (see Figure 8.4). This example also illustrates the importance of good visualization to be able to see correlations in the data and determine important features. Furthermore, it shows once again the elegance of boxplots that contain a lot of information.

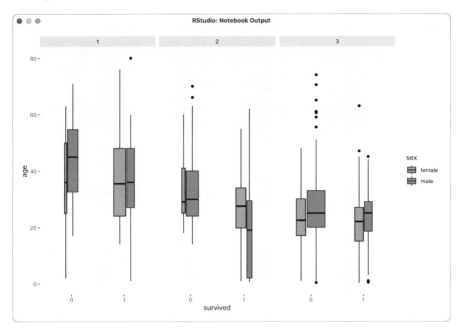

FIGURE 8.4
Boxplots with "facet" passenger class and the variables age and gender.

The other variables we haven't included yet include some that probably won't have an impact, such as the ticket number, but, leave it to the reader to try everything out for themselves.

8.2.2 The Caret Package: Dummy Variables and Splitting the Data

We have a little over 1,300 data points – not a lot, but at least in this case, enough to try out different algorithms. First, however, we need to divide our data into training data and test data (see Chapter 3, "The Anatomy of a Data Science Project"). This is where the Caret package helps us. Caret stands for *Classification And Regression Training*, and the package offers many functions that will make our lives easier. One of these functions is the creation of dummy variables. As described in Chapter 3, "Flow of a Data Science Project," we need dummy variables for some algorithms when we have non-numeric data but numeric data is needed for distances. We should do this before deciding how to split the data. We first load the library caret and then create the dummy variable object dmy:

```
> library(caret)
> dmy <- dummyVars(" ~ sex", data = titanic, fullRank=T)
```

The parameter fullRank=T ensures that we have one fewer column than there are categories. In our example, there are two possibilities in the variable sex, male or female, so we can save one of them:

```
> trsf <- data.frame(predict(dmy, newdata = titanic))
 > head(trsf)
  sexmale
1 0
2 1
3 0
4 1
5 0
6 1
>
```

Additionally, we need to modify the data somewhat. Based on the short data analysis, we are interested in the variables passenger class and age in addition to gender. However, age is imported as a character, and decimal places are separated by a comma, which R does not understand. We could have fixed this during the import, but it is not uncommon to notice this after the data has been loaded. We have to convert this as well. And finally, we make a factor out of our target variable, survived:

```
titanic_subset <- titanic %>%
    select(survived,pclass,age) %>%
    mutate(age = str_replace(age,",".")) %>%
    mutate(age = as.double(age)) %>%
    mutate(survived = as.factor(survived))
```

Now, the data set and the dummy variable are merged and NAs are removed:

```
titanic_subset <- cbind(titanic_subset,trsf)
titanic_subset <- titanic_subset %>%
filter(!is.na(age))
```

Now we can start dividing the data into training and test data. The special thing here is that the caret package also takes care that the two sets of data are balanced. This will be explained in more detail.

Assuming that we take a sample of 340 elements from the 1300 rows (i.e. about 20%) for a later test, and leave the remaining 80% or so for training, it would be desirable to have the same ratio of survivors and non-survivors in the two subsets. However, it is equally important that we also have the same ratio of passenger classes, sex, and age, because suppose by chance we had many more males in one subset; this would affect the performance of our model. This is exactly what the caret package takes care of;

the function createDataPartition() takes care of the balance of the feature attributes:

```
set.seed(456)
trainIndex=createDataPartition(titanic_subset$survived,
    p=0.8)$Resample1
train.data=titanic_subset[trainIndex, ]
test.data=titanic_subset[-trainIndex, ]
```

We now check what the ratios of the classes and the variables sex look like:

```
> table(titanic$survived,titanic$sex)

    female male
0 127 682
1 339 161
>
> table(train.data$survived,train.data$sex)

      0 1
0 85 411
1 231 111
>
```

Obviously, the caret package has done a good job here.

8.2.3   The pROC Package

Before we get into the algorithms, let's look at how we want to measure their performance. In Chapter 3, "The Anatomy of a Data Science Project", we already covered the ROC AUC, here, we will use the package pROC:

```
library(pROC)
```

We will plot additional information in the following examples, namely a confidence interval. More about this in the individual sections on the algorithms.

8.3   Decision Trees

We come to our first classification algorithm, decision trees. First approaches to this algorithm appeared in the 1950s, and today, decision trees are still popular because their results are comprehensible and relatively simple, which also increases their acceptance by users.

8.3.1   How the Algorithm Works

A decision tree grows from the top down. At the top-lying root, the data is, in simple terms, divided into two parts in such a way that the largest possible difference in the target variable is created. Explained using our Titanic data set as an example, the variable that best splits the data in terms of the target variable survived/non-survived is used first. Based on our short EDA, this will likely be gender. This step is then repeated until further splitting no longer improves the results. When we reach the end of the branches at the leaves we can read a result. This procedure is also called recursive partitioning.

8.3.2   Training and Test

We use two libraries for our decision tree; on the one hand, rpart for the creation of the tree, and on the other hand, rpart.plot for the visualization:

```
library(rpart)
library(rpart.plot)
```

We start by training the model, for this, we use the training data:

```
tree<- rpart(survived~., data = train.data, method = 'class')
```

The code requires some explanation. Like almost all classification algorithms, rpart() first requires a formula. This has the form

```
Dependent variable ~ Independent variable
```

In this case, *survived* is the dependent variable, the dot after the tilde stands for all other columns. We also specify the data set (data =) and the method we want to use, in this case a classification. Now we plot the model we just created:

```
rpart.plot(tree)
```

The result can be seen in Figure 8.5. This could be understood as a "set of rules" – and this is also the reason why users trust this approach, because it allows them to understand exactly what the algorithm does.

As expected, gender is the variable that makes the most difference at the beginning and is therefore used at the root. At each node are the predicted class, the predicted probability of surviving, and the percentage of observations in the node. So at the top, we are told that the probability of surviving is 41%. If we go to the right to females, the probability of surviving is 73%; 38% of the cases match this attribute. Now, we branch out further: If the passenger class is less than 3 (we continue to the right, which is the path for "no"), then the probability of surviving is 92%. However, if the passenger class is third class, then the probability is only 46%. Then, there is further subdivision: If

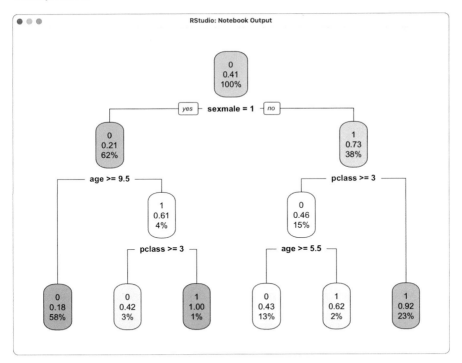

FIGURE 8.5
The result of our first decision tree.

the age is greater than/equal to 5.5 years, then the probability of surviving is
still 43%; if the passenger is younger, 62%. Quite simple, really, isn't it?

Now, let's throw this model at the test data and then, let's output the first
lines of results with the data:

```
> dt_results <- predict(tree, test.data[,-1], type = 'prob')
> head(model.results.dt <- cbind(test.data,dt_results))
   survived pclass age sexmale 0 1
7  1 63 0 0.08465608 0.9153439
11 0 1 47 1 0.81687243 0.1831276
17 0 1 24 1 0.81687243 0.1831276
22 1 1 47 0 0.08465608 0.9153439
24 1 1 42 0 0.
08465608 0.9153439
35 0 1 42 1 0.81687243 0.1831276
```

We convert the probabilities to ones and zeros to be able to create a diffusion matrix:

```
test.results <- model.results.dt %>%
mutate(pred = ifelse('1'>=0.5,1,0))
 > table(test.results$pred, test.data$survived)

    0 1
0 122 33
1 1 52
>
```

8.3.3   Interpretation of the Results

The diffusion matrix can be read as follows:

- 122 of the non-survivors were correctly classified as such.

- 1 non-survivor was misclassified as a survivor.

- 33 survivors were misclassified as non-survivors.

- 52 survivors were correctly classified as such.

The decision tree does a very good job on the non-survivors, but the survivors are a different story; 19 of the 33 misclassified survivors were actually in the third passenger class, and their probability of surviving was very low. Nevertheless, the fact that they were male and over 9.5 years old was enough to classify them as non-survivors. We create the ROC-AUC graphic; you can read about the individual parameters in the package help:

```
pROC_obj <- roc(model.results.dt$survived,model.results.dt$'1',
smoothed = TRUE,
ci=TRUE, ci.alpha=0.9, stratified=FALSE
, plot=TRUE, auc.polygon=TRUE, max.auc.polygon=TRUE, grid=TRUE,
print.auc=TRUE, show.thres=TRUE)
sens.ci <- ci.se(pROC_obj)
plot(sens.ci, type="shape", col="lightblue")
plot(sens.ci, type="bars")
```

The result in Figure 8.6 is striking because we see a completely straight line upwards. If we look at the confusion matrix, this is relatively easy to explain, as we have only one misclassification in the group of non-survivors.

0.854 is already a pretty good value, and we haven't optimized anything yet. Let's see how the other algorithms will perform.

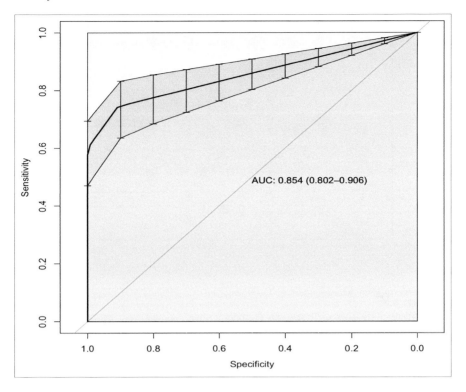

FIGURE 8.6
ROC AUC for our decision tree model.

8.4 Support Vector Machines

This algorithm from the 1950s (Ronald A. Fischer had the first thoughts in the direction of a SVM already in the 1930s) offers a very elegant way to classify data points. It is used, among other things, for the ranking of results in search engines (Ranking SVMs).

8.4.1 How the Algorithm Works

The principle of support vector machines is again based on distances. In a set of points, an attempt is made to draw a straight line, quite analogous to linear regression. Instead of a line, which could be drawn at different angles through the points, here, parallel lines are drawn around the line, the so-called support vectors. And now, as in linear regression, several lines are tried, and for each, it is calculated how far the support vectors can be from the line without touching any points. The goal is to get the support vectors as far

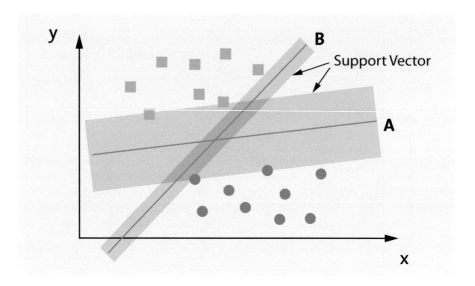

FIGURE 8.7
Two different lines A and B with their support vectors (graphic by F. Bürger, sh. Wikipedia keyword "Support Vector Machine", own addition).

away from the line as possible. The line with the support vectors furthest away is the ideal line.

Let's take a look at the example in Figure 8.7. Line B passes through the points in such a way that the support vectors are relatively close to the line. With line B, the line is drawn through the points in such a way that the parallel lines can be further away. Line A is therefore a better solution for the algorithm.

The data will not always be as clearly separated as in the example. For example, angular data points may lie with the circular data points, and vice versa. In this case, the procedure is still the same as just described, except that for each point that lies on the wrong side, an error is calculated based on how far into the wrong area it is. The more errors, the worse the support vectors lie, and the greater the penalty, so to speak. This approach to minimize this penalty is called the *loss function*. In this case, the SVM that has the smallest error is searched for.

Now, we have more than two dimensions, in this case we use three. This case is still relatively easy to understand, because we can now draw in a *plane* that separates the data points in the three-dimensional space. For higher dimensions, i.e. more than three, the procedure works analogously, even though it can no longer be visualized in space. For three dimensions and more, the plane that passes through points is also called a *hyperplane*.

If we look at our data points in Figure 8.8, we can already see how such a hyperplane could go through the data points. The data is not scaled yet;

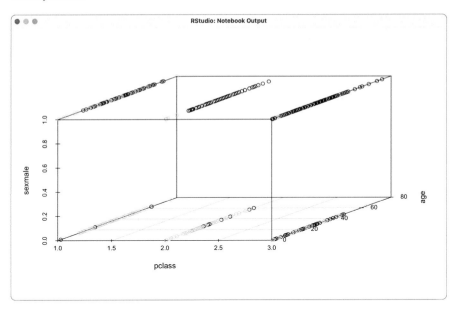

FIGURE 8.8
Our data points plotted; light gray is for survivors, black for non-survivors.

our SVM function will do that afterwards. For more dimensions, we can't plot it, but even if it is hard to imagine for us humans, purely mathematically it would work. Spaces with more than three dimensions are also called hyperspaces. But what if no straight line or hyperplane can be drawn through the data, because they are perhaps separated in a completely different way?

SVMs can help here as well. For example, data can be easily transposed into a higher-dimensional space, and "cuts" of the resulting shape can be used there, using the so-called kernel trick. A kernel is a mathematical approach in machine learning to perform computations in higher dimensions.

An example of this can be seen in Figure 8.9. The data are initially in a two-dimensional space. By adding a third dimension, there is a way to separate them linearly, that is, to run a straight plane through them. The package e1071, which we will use, offers besides the linear kernel also a polynomial, a radial basis function (RDF), and a sigmoid-function. The RDF kernel can be applied when you do not have any information about the data yet. The sigmoid kernel is commonly used in neural networks, and the polynomial kernel is a generalized variant of the linear kernel. The choice of a kernel is of immense importance for the performance of a support vector machine.

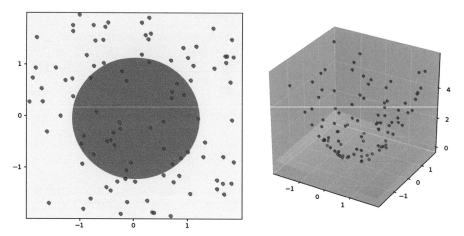

FIGURE 8.9
Kernel trick by adding a dimension. Wikipedia, CC BY-SA 4.0, by Shiyu J.

8.4.2   Data Preparation

We can reuse the training and testing data from the decision tree section here without customizing it.

8.4.3   Training and Test

For our SVM, we use the package e1071. The name comes from a department of the Vienna University of Technology, which has provided several statistical functions in this package. After that, we train the classifier:

```
library(e1071)
classifier = svm(survived ~ .,
data = train.data,
type = 'C-classification',
probability = TRUE,
kernel = 'linear')
```

Let's take a closer look at the individual parameters. You have already got to know the format of the formula with the decision trees also that the data is defined. Next, we determine what we want to do with the support vector machine, in this case a classification. We also want to output the probabilities and choose the parameter probability = TRUE for this. Finally, we define the kernel, in this case a linear kernel. By the way, we did not scale the data, because the svm() function scales the data automatically. Now, we cast the resulting model onto the test data:

```
> svm.predict = predict(classifier, newdata = test.data[,-1],
```

```
probability = TRUE)
```

Now let's look at the results first:

```
> table(test.data$survived, svm.predict) # Confusion matrix
svm.predict
      0 1
0 112 11
1 24 61
>
```

Obviously, the SVM works better for the survivors that the decision tree had classified as non-survivors. At the same time, however, it looks worse at classifying the non-survivors.

8.4.4 Interpretations of the Results

We take a look at the first results:

```
> head(cbind(test.data,svm.predict), 10)
   survived pclass age sexmale svm.predict
7  1 1 63 0 1
11 0 1 47 1 0
17 0 1 24 1 0
22 1 1 47 0 1
24 1 1 42 0 1
35 0 1 42 1 0
43 1 1 59 0 1
50 1 1 36 1 0
51 1 1 58 0 1
53 0 1 28 1 0
>
```

We don't see anything conspicuous at first; there is also only one misclassification to be seen in row 50, where the SVM has predicted that the passenger will not survive. Since he is a man and also no longer a child, we can already guess what this might be due to. After that, we plot the ROC AUC:

```
probs <- attr(svm.predict, "prob")[,2]
pROC_obj <- roc(as.numeric(test.data$survived), probs,
smoothed = TRUE,
ci=TRUE, ci.alpha=0.9, stratified=FALSE,
plot=TRUE, auc.polygon=TRUE, max.auc.polygon=TRUE, grid=TRUE,
print.auc=TRUE, show.thres=TRUE)
sens.ci <- ci.se(pROC_obj)
plot(sens.ci, type="shape", col="lightblue")
plot(sens.ci, type="bars")
```

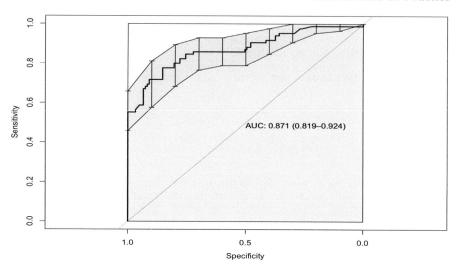

FIGURE 8.10
ROC-AUC diagram of the SVM.

You can see the result in Figure 8.10. We have a slightly better AUC value compared to the decision tree. Also, in this plot, the ROC curve goes steeply up; likewise, the curve is very straight at specificity starting at about 0.25. This is consistent with the observations in the confusion matrix.

8.5 Naive Bayes

The Naive Bayes algorithm is based on Bayes' Theorem, named after the mathematician and priest Thomas Bayes, who lived in the 18th century. His mathematical theorem describes the calculation of *conditional probabilities* and was published only after Bayes' death with the essay "An essay towards solving a problem in the doctrine of chances".

To understand conditional probabilities, let's first look at a simple example from probability theory, the roll of a die. Please don't be put off by the formulas that follow, you'll see it's not that hard!

The probability that we roll a 6 – let's call this event A – is described by the expression $p(A)$, p for probability. If the dice are fair, the probability of rolling a 6 is 1/6. If we roll the dice again afterwards – let's call this event B – the probability is 1/6 again, because the result of the second roll is independent of the result of the first roll. Just because I had a 6 on the first roll, the probability of a 6 on the second roll does not change. Or, in other words, the result of the first throw does not give us any information about what the result of the second throw will be.

The probability that event A and event B occur together is described by the expression $P(AB)$:

$$P(AB) = P(A) \cdot P(B)$$

Using the example of a die: The probability that another 6 is rolled after a 6 is

$$P(AB) = \frac{1}{6} \cdot \frac{1}{6} = \frac{1}{36}$$

This is also the probability of first rolling a 6 and then a 1. Events are not always independent of each other. The probability of encountering a man with red-green vision impairment is greater than that of encountering a woman with such impairment, since men suffer from it more frequently.

Back to our dice experiment. Suppose I have six dice in my pocket, all of them are marked, because the first die has only 1s on all sides, the second has only 2s, and so on. In this case, the rolls are not independent of each other, because if I roll a 6 with one of these dice, then there will also be a 6 on the rolls after it.

Since there is a probability of 1/6 that I got the die with the 6s out of my pocket, and after that, the result of the first roll determines the result of all other rolls, the events are no longer independent of each other. We could also express it like this:

$$P(AB) = P(A) \cdot P(A|B)$$

The part $P(B|A)$ can be expressed colloquially like this: If we know A, what is the probability of B? If the die is fair, then roll A tells us nothing about roll B. If the die is not fair, then A tells us a great deal about the outcome of B. And it is this case that corresponds to a conditional probability. The prior knowledge is also called a *prior*.

By some mathematical transformations, we can write this formula like this:

$$P(B|A) = \frac{P(B) \cdot P(A|B)}{P(A}$$

And this is exactly the theorem of Bayes.

8.5.1 How the Algorithm Works

We now want to use Bayes' theorem for a classification and predict whether or not a person on the Titanic died based on their feature attributes. To do this, we take Bayes' theorem, but replace A and B with E and c – E for evidence and c for class (died or not died, not passenger class):

$$P(c|E) = \frac{P(c) \cdot P(E|c}{P(E)}$$

In other words, what is the probability that someone belongs to the "survived" class given evidence? In our Titanic example, we have several conditions: age, gender, and passenger class. But first, let's look at what the formula for a condition in Bayes' theorem would look like:

$$P(survived|male) = \frac{P(male|survived) \cdot P(survived)}{P(make}$$

If we now have more than one piece of evidence, we can also write:

$$P(c|E) = \frac{P(e_1|c) \cdot P(e_2|c) \cdot ... \cdot P(e_n|c) \cdot P(survived)}{P(E)}$$

For p(E) holds:

$$P(E) = P(e_1|c_0) \cdot P(e_2|c_0) \cdot ... \cdot P(e_n|c_0) \cdot P(c_0)$$
$$+ P(e_1|c_1) \cdot P(e_2|c_1) \cdot ... \cdot P(e_n|c_1) \cdot P(c_1)$$

Using a single evidence example:

$$P(E) = P(E|survived) \cdot P(survived) + P(|notsurvived) \cdot P(notsurvived)$$

Now, we have covered the details of the Naive Bayes classifier. It calculates the frequencies of classes and features and how they appear together, and every bit of information is taken into account.

What is naive about Naive Bayes?

Naive Bayes is called *naive* because it assumes the independence of events. This is not always true; however, it works nevertheless. If you look at some spam mails, it is hard to imagine that some terms would appear independently.

From now on, the Naive Bayes classifier will take over these computational steps for us. This algorithm is exciting in any case, since it does not require any distances, but is based on a completely different approach and gets by with relatively few resources.

8.5.2 Data Preparation

We will use the library e1071 again. It can handle a data frame with categorical as well as numerical data. Nevertheless, we convert the age by classifying all passengers over 14 as adults, and also convert the passenger class into a factor:

```
train.data.nb <- train.data %>%
    mutate(pclass = as.factor(pclass)) %>%
    mutate(age = as.factor(if_else(age > 14, "adult", "child")))

data.nb <- test.data %>%
    mutate(pclass = as.factor(pclass)) %>%
    mutate(age = as.factor(if_else(age > 14, "adult", "child")))
```

The choice of buckets (we could have chosen more, for example, infant, child, teenager, adult, senior) can be of great importance, because it divides into more or fewer classes, which in turn, can be an advantage to the prediction.

8.5.3 Training and Test

We use the naiveBayes() function and output the result of the model right away:

```
> Naive_Bayes_Model=naiveBayes(survived ~., data=train.data.nb,
    type = "raw")
> Naive_Bayes_Model

  Naive Bayes Classifier for Discrete Predictors
Call:
naiveBayes.default(x = X, y = Y, laplace = laplace, type = "raw")
A-priori probabilities:
Y
0 1
0.5918854 0.4081146
Conditional probabilities:
pclass
Y 1 2 3 0
0.1612903 0.2358871 0.6028226
1 0.4122807 0.2690058 0.3187135
age
Y adult child
0 0.9233871 0.0766129
1 0.8538012 0.1461988
sexmale
```

```
Y [,1] [,2] 0 0.
8286290 0.3772132 1 0.
3245614 0.4688968
>
```

We see here the *a priori* probabilities, namely, the probability that a passenger survives or not, and below that, the conditional probabilities for each variable. The more features we have, the less we will be able to look at all the data. We apply the model to the test data:

```
> Titanic.class <- predict(Naive_Bayes_Model, test.data.nb[,-1])
> table(Titanic.class, test.data.nb$survived)

 Titanic.class 0 1
0 112 24 1
11 61
>
```

The results are similar to those of SVM. We do not seem to have won much.

8.5.4   Interpretation of the Results

We need to re-run our classifier again on the test data to get the probabilities. Then we create the ROC-AUC-graph:

```
Titanic.class <- predict(Naive_Bayes_Model, test.data.nb[,-1],
    type = "raw")
titanic.probs <- Titanic.class[,2]
pROC_obj <- roc(as.numeric(model.results$survived),
titanic.probs,
smoothed = TRUE,
ci=TRUE, ci.alpha=0.9, stratified=FALSE
, plot=TRUE, auc.polygon=TRUE, max.auc.polygon=TRUE, grid=TRUE,
print.auc=TRUE, show.thres=TRUE)
sens.ci <- ci.se(pROC_obj)
plot(sens.ci, type="shape", col="lightblue")
plot(sens.ci, type="bars")
```

The result can be seen in Figure 8.11. Again, no surprises with respect the confusion matrix. The result is only minimally better.

8.6   XG Boost: The Newcomer

So far, you've encountered many algorithms that had their origins decades ago, and in some cases, even centuries ago. XG Boost is not completely without

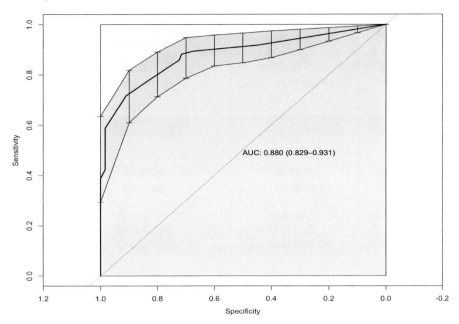

FIGURE 8.11
ROC-AUC diagram for our Naive Bayes classifier.

history either, as you will see in a moment, but the underlying paper was only published in 2016 [16]. Since then, XG Boost has achieved considerable success in many Data Science projects.

8.6.1 How the algorithm works

XG Boost stands for eXtreme Gradient Boost(ing), which introduces two new terms, gradient and boosting. Before we get into the theory, let's tie in with what you already know. At the beginning of this chapter, you learned about decision trees. Let's imagine that we have not just one tree, but several trees. In machine learning language, this is called an ensemble. Figure 8.12 shows how an ensemble works.

Such an ensemble might consist of different models, each with different strengths and weaknesses, and the correct predictions of the models might each reinforce each other, while the incorrect ones cancel each other out. An example of this is random forrests. In boosting, which is an approach to ensemble learning, the models learn sequentially, with each model learning from the "mistakes" of the previous model. The gradient is used to optimize loss functions which you have already learned about in support vector machines. That is, each model tries to minimize the errors of the previous model.

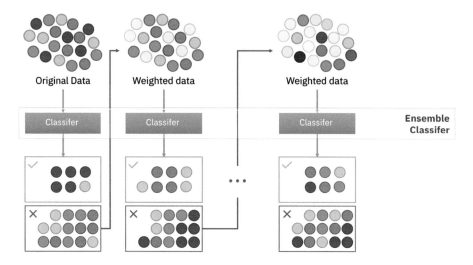

FIGURE 8.12
Ensemble learning visualized (from: Wikipedia, CC BY-SA 4.0, by Sirakom).

8.6.2   Data Preparation

First, we load the library xgboost that must be installed before loading (see Chapter 4, "Introduction to R"):

```
library(xgboost)
```

Then, we create a matrix from our data frame, since the xgboost() function expects a matrix. Additionally, we modify our target variable:

```
dtrain <- as.matrix(train.data[,-1])
mode(dtrain) <- 'double'
train.data.xg <- train.data %>%
mutate(survived = as.integer(if_else(survived == "0",0,1)))
Now we are ready for the training.
```

8.6.3   Training and Test

The xgboost() function takes a few more parameters than we are used to from the previous examples. We will start the training, and then, I will explain the individual parameters:

```
xgb <- xgboost(
dtrain
,
 label = train.data.xg$survived,
nrounds=100,
```

```
eval_metric = "error",
max_depth = 10,
eta = 0.1,
verbose = 0,
nthread =
8,
objective = "binary:logistic"
)
```

What do the parameters mean?

- nrounds is the number of boosting iterations.

- eval_metric allows us to observe other metrics, in this case, logloss and error.

- max_depth is the maximum depth of a tree.

- eta is the learning rate whose adaptation can prevent overfitting.

- verbose determines in how much detail xgboost messages are shown; 0 means that XG Boost is mute.

- nthread is the number of parallel threads.

- objective is the goal, in this case, binary:logistic.

We start to classify with our model:

```
xg_matrix <- as.matrix(test.data[,-1])
pred <- predict(xgb, xg_matrix)
print(head(pred))
```

```
[1] 0.82253844 0.05645552 0.38106889 0.96243805 0.98194110
    0.67908919
```

To create a confusion matrix, we can either convert the values in pred to 1 or 0 (as we have already done earlier), or use the function confusionMatrix from the caret package. But for this, we need a 1 or 0 from the classification and no probability:

```
xgb2 <- xgboost(
dtrain
,
 label = train.data.xg$survived,
nrounds=100,
eval_metric = "error",
max_depth = 10,
eta = 0.1,
verbose = 0,
```

```
nthread =
8,
objective = "binary:hinge"
)
pred2 <- predict(xgb2, xg_matrix)
```

Now we can use the function:

```
> confusionMatrix(as.factor(pred2),factor(test.data$survived))
Confusion Matrix and Statistics
Reference
Prediction 0 1
0 112 21
1 11 64


                Accuracy : 0.8462
95% CI : (0.7898, 0.8923)
No Information Rate : 0.5913
P-Value [Acc > NIR] : 1.596e-15


                   Kappa : 0.6758

  Mcnemar's Test P-Value : 0.1116


             Sensitivity : 0.9106
Specificity : 0.7529
Pos Pred Value : 0.8421
Neg Pred Value : 0.8533
Prevalence : 0.5913
Detection Rate : 0.5385
Detection Prevalence : 0.6394
Balanced Accuracy : 0.8318

        Positive' Class : 0

 >
```

The values explained:

- The Accuracy is the percentage of observations that were correctly classified.

- 95% CI stands for the 95% confidence interval.

- No Information Rate: Approximately 59 % of the passengers did not survive, so that a classifier would already be 59% correct if it simply grouped all passengers as not surviving. This value must therefore be taken into account when looking at the accuracy.

- I explain the P-Value in the Appendix.

- Kappa: A special statistical test; in this case, we look at how much better the classifier is than random values.

- Mcnemar's Test P-Value: McNemar's Test is a special statistical test for paired data.

I have already explained specificity and sensitivity. The remaining values are not important for us for the time being.

8.6.4 Interpretation of the Results

We'll take a look at the results first:

```
> xgb.df <- as.data.frame(pred)
> head(results.xg2 <- cbind(xgb.df,test.data))
          pred survived pclass age sexmale
7  0.82253844 1 1 63 0
11 0.05645552 0 1 47 1
17 0.38106889 0 1 24 1
22 0.96243805 1 1 47 0
24 0.98194110 1 1 42 0
35 0.67908919 0 1 42 1
>
```

Now, we create the ROC-AUC diagram:

```
pROC_obj <- roc(results.xg2$survived,results.xg2$pred,
smoothed = TRUE,
ci=TRUE, ci.alpha=0.9, stratified=FALSE,
plot=TRUE, auc.polygon=TRUE, max.
auc.polygon=TRUE, grid=TRUE,
print.auc=TRUE, show.thres=TRUE)
sens.ci <- ci.se(pROC_obj)
plot(sens.ci, type="shape", col="lightblue")
plot(sens.ci, type="bars")
```

You can see the result in Figure 8.13.

Obviously, XG Boost has done a pretty good job here; the algorithm has already achieved the best value compared with all previous models without any optimizations. In addition, the package xgboost contains the function xgb.importance(), which can be used to display the importance of the individual features:

```
> mat <- xgb.importance (feature_names =
```

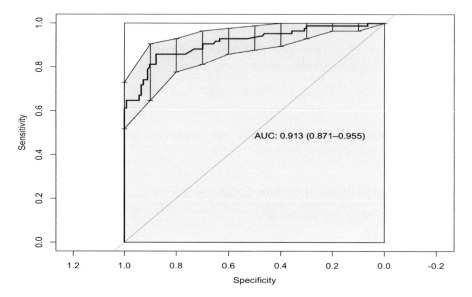

FIGURE 8.13
ROC-AUC diagram for our XG-Boost model.

```
colnames(train.data.xg[,-1]),model = xgb)
 > mat
   Feature Gain Cover Frequency
1: sexmale 0.4617308 0.09938245 0.03265857
2: age 0.3096015 0.77798739 0.83805668
3: pclass 0.2286677 0.12263016 0.12928475
>
```

The three columns are explained as follows:

- The gain is the contribution of each feature to the model.

- The cover is the number of observations related to this feature.

- The frequency specifies how often the feature appears in the trees.

  In addition, we can plot the result, where only the gain is shown here:

```
> xgb.plot.importance (importance_matrix = mat[1:10])
```

The result of the plot can be found in Figure 8.14. As with our first decision tree, the gender also plays the most important role here. However, we don't see the "rules" here, which immediately makes the algorithm a bit more opaque to decision tree followers. It is helpful to explain that there is a similar system at work here, except that there is more than one tree.

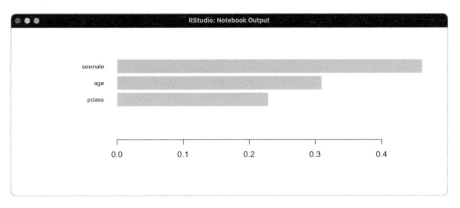

FIGURE 8.14
Plotted importance table of XG Boost.

8.7 Text Classification

Spam is unfortunately a ubiquitous phenomenon, so why not try to build a classifier that can distinguish spam from legitimate mail? In this section, we once again use the Naive Bayes-classifier, which has generally shown good results in blocking spam. Feel free to try the other algorithms as well; you will acquire the necessary tools in the following sections.

I will not describe again how Naive Bayes works, but what is new in this section is how text is handled in general. By the way, modern spam classifiers do not work on the basis of content alone; nowadays many other signals are included, such as the sender address or certain server settings.

8.7.1 Prepare the Data

As a data set, we use a collection of emails that are already classified. You can find this data set at https://alby.link/spam or on the website for the book. It is best to load the data into a directory data, and then read it in:

```
> emails = read.csv('data/emails.csv', stringsAsFactors = FALSE)
> table(emails$spam)

   0 1
4360 1368
>
```

We have 1,368 spam emails and 4,360 normal emails. You can get a first impression in Figure 8.15. We use two libraries: tm, which is used for text mining, and SnowballC, a stemer. To explain this in more detail: Words can

text	spam
Subject: naturally irresistible your corporate identity I...	1
Subject: the stock trading gunslinger fanny is merrill ...	1
Subject: unbelievable new homes made easy im want...	1
Subject: 4 color printing special request additional in...	1
Subject: do not have money , get software cds from h...	1
Subject: great nnews hello , welcome to medzonline ...	1
Subject: here ' s a hot play in motion homeland secur...	1
Subject: save your money buy getting this thing here ...	1
Subject: undeliverable : home based business for gro...	1
Subject: save your money buy getting this thing here ...	1
Subject: las vegas high rise boom las vegas is fast be...	1
Subject: save your money buy getting this thing here ...	1

FIGURE 8.15
The first lines of the spam record.

be capitalized or not, they are changed ("one dream, several dreams"), and many words have no important meaning (e.g., "the", "a", etc., so-called stop words). We therefore transform the words in our data frame:

- We convert everything to lower case.

- We convert everything to "Plain Text".

- We remove punctuation.

- We remove the stop words.

- We extract the root word (i.e. "dream" instead of "dreams").

We store all this in a corpus, a collection of texts for linguistic analysis. Once in code form:

```
library(tm)
library(SnowballC)
corpus = VCorpus(VectorSource(emails$text))
```

```
corpus = tm_map(corpus, content_transformer(tolower))
corpus = tm_map(corpus, PlainTextDocument)
corpus = tm_map(corpus, removePunctuation)
corpus = tm_map(corpus, removeWords, stopwords("en"))
corpus = tm_map(corpus, stemDocument)
```

Next, a *document term matrix*, which is nothing more than a matrix in which each word has its own column, and a 1 or 0 indicates whether it occurs in the document rows:

```
> dtm = DocumentTermMatrix(corpus)
> dtm
<<DocumentTermMatrix (documents: 5728, terms: 28687)>>
Non-/sparse entries: 481719/163837417
Sparsity : 100%
Maximum term length: 24
Weighting : term frequency (tf)
>
```

However, we don't want to keep all words in the document term matrix, because some words occur very rarely. Therefore we remove rare words with the function removeSparseTerms(). Unfortunately, the documentation of this function is also "sparse". Let's take a look at what it does:

```
> spdtm = removeSparseTerms(dtm, 0.95)
> spdtm
<<DocumentTermMatrix (documents: 5728, terms: 330)>>
Non-/sparse entries: 213551/1676689
Sparsity : 89%
Maximum term length: 10
Weighting : term frequency (tf)
>
```

Now, we have only 330 words instead of 28,678. The sparsity has dropped from 100% to 89%. Most of the words appear in only a few documents (except for the stop words, which we have already removed). If the value for sparse is chosen very low, then we have only words that appear in almost all documents. If the value is chosen higher, we have more documents in which not all words occur. This has advantages and also disadvantages. The advantage is that our matrix is much smaller and easier to manage. The disadvantage is that some words occur very rarely, but still have an important meaning for this document. In our example, however, this is probably less important, since we are looking for words that can serve as a signal for spam. Let's take the following spam email:

```
Hello %address_e-mail%,
A message awaits you in your customer area. It informs you about
```

upcoming changes to your account agreement.
We invite you to read it by logging in by clicking here
Thank you for your attention.
Respectfully,
Online Customer Service.

Apart from the fact that the spammer didn't put much effort into it, we can see that it makes sense to have not only the most popular words in the matrix ("we" will probably appear in many documents); however, we still want to keep such words as "customer area". The definition of a sparse value here can have a big impact on the performance of the spam classifier.

In the next step, we create a data frame:

```
emailsSparse = as.data.frame(as.matrix(spdtm))
colnames(emailsSparse) = make.names(colnames(emailsSparse))
emailsSparse$spam = emails$spam
emailsSparse$spam = as.factor(emailsSparse$spam)
```

From the document term matrix we first had to build a "normal" matrix before converting it to a data frame. Then, we add the target variable and change it to a factor.

8.7.2 Training and Test

Now, as before, the corpus is divided into training and test:

```
set.seed(983)
trainIndex=createDataPartition(emailsSparse$spam, p=0.8)
    $Resample1
train.data=emailsSparse[trainIndex, ]
test.data=emailsSparse[-trainIndex, ]
```

As a next step, a Naive Bayes model is formed, which we immediately use with our data:

```
> Naive_Bayes_Model=naiveBayes(spam ~., data=train.data,
    type = "raw")
> Spam.class <- predict(Naive_Bayes_Model, test.data)
> table(Spam.class, test.data$spam)

 Spam.class 0 1
0 707 9
1 165 264
>
```

On the one hand, we have few misclassifications: Nine emails that were not spam were classified as spam, but at the same time, a relatively large number of emails were not detected as spam.

8.7.3 Interpretation of the results

We still need to transform our results a bit, because we need the probabilities, so we create a new prediction with the test data:

```
model.results <- cbind(test.data,Spam.class)
Spam.class <- predict(Naive_Bayes_Model, test.data, type = "raw")
real_spam <- as.numeric(as.character(model.results$spam))
```

Now, we create the ROC-AUC diagram:

```
pROC_obj <- roc(real_spam, Spam.class[,2],
smoothed = TRUE,
          ci=TRUE, ci.alpha=0.9, stratified=FALSE,
plot=TRUE, auc.polygon=TRUE, max.auc.polygon=TRUE, grid=TRUE,
print.auc=TRUE, show.thres=TRUE)
sens.ci <- ci.se(pROC_obj)
plot(sens.ci, type="shape", col="lightblue") plot
(sens.ci, type="bars")
```

The result can be seen in Figure 8.16.

The picture from our confusion matrix is also shown here in the ROC-AUC diagram. We have a very straight line from a specificity of 0.7. At the same time, we have managed exactly what is desirable: as few false positives as possible (mails that are not spam but are classified as such), even if this means accepting more false negatives.

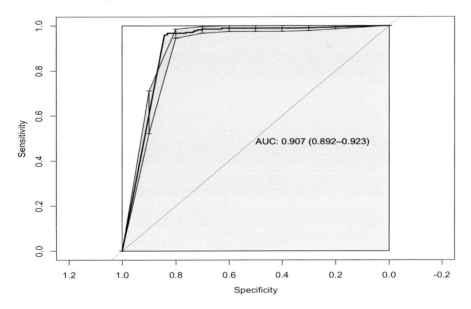

FIGURE 8.16
ROC-AUC diagram of our spam classifier.

9

Other Use Cases

Machine learning is not just about prediction, clustering, or classification. For example, how can an algorithm suggest a relevant song to a music lover? How do we identify *proximity* of objects? That's what this chapter is about.

Recommending relevant products is the supreme discipline for many marketing people, and those who love music are always on the lookout for the next song, the next record. We used to browse through friends' record or CD collections or use the radio to discover new music. Today, this task is taken over by algorithms.

9.1 Shopping Cart Analysis – Association Rules

If you buy apples, you also put potatoes in your shopping basket – this type of data mining is based on association rules, which are not only common in e-commerce. Association rules can be used to make relevant recommendations, whether for additional products that could be added to a shopping cart or for content that matches what the user has just read.

9.1.1 How the Algorithm Works

A common question in this context is why not simply display the items that are most popular, i.e. most frequently purchased? This will be explained using the example of a music service. At the time of writing this manuscript, both Ed Sheeran and Helene Fischer are in the top 10 of the German singles charts. Of course, both artists deserve respect for the fact that they seem to have released very successful songs, but that won't entice me to listen to their music. Spotify, on the other hand, usually does a good job of suggesting music that I like. This is done based on what I've listened to so far, which is then compared with playlists that contain my tracks and have a lot of followers at the same time.

Usually, nothing from the current charts is on there. When Spotify recommends something to me, it works thanks to optimized rules that say exactly how likely I am to like a track. You could now argue that I have a very specific taste in music if I don't like the charts.

DOI: 10.1201/9781003426363-9

But, imagine you have to go to the supermarket and you see an ad there that you should buy milk, toilet paper, and butter, due to the fact that these are what is bought most often. Just because it's common, this doesn't mean it's relevant to everyone. In addition, music lovers in particular are all about discovering something new. This will hardly work with the most frequently heard songs.

Association search has proven to be very successful in the creation of recommendations. However, it can be "expensive" in terms of its computation performance because, in principle, the occurrence of each element must be identified with that of another. Based on a paper by [2] which focused on data mining of large data sets, the Apriori algorithm emerged, which we will also use in this chapter. This algorithm is a special approach that not only ignores rarely occurring items, but also ignores the transactions in which those rare item appear. This reduces the amount of data to be computed.

Let's imagine that beer is purchased in 10% of all purchases and potato chips in 7% of purchases. If we only look at the beer purchases and only 7% of these purchases include potato chips, then there is no correlation. However, if 25% of all beer purchases also include potato chips, then there seems to be an association here. An association is not limited to two products; for example, beer may be associated with potato chips, wine, and cheese cubes. These associations need not always be strong.

The package arules contains an implementation of the Apriori algorithm. We first start with a very simple, self-constructed example, so that you can understand all the calculations.

9.1.2 Data Preparation

Let's now look at the previous example of purchasing in detail. First, we install the package arules and then, load the library:

```
> install.packages("arules")
> library(arules)
```

For the first example, we'll create a simple data set of shopping baskets to make the results easier to follow; you can find this data set on the website for the book, so you don't have to type it out. The shopping baskets:

```
a_list <- list(
# The sample data, each line here is a receipt
c("
apple", "beer", "rice", "chicken"),
c("apple", "beer", "
rice"),
c("apple", "beer"),
c("apple", "pear"),
c("milk", "beer", "rice", "chicken"),
```

```
c("milk", "beer", "rice"),
c("milk", "beer"),
c("apple", "pear") )
```

The data must now be converted to a special format:

```
trans <- as(a_list, "transactions")
```

If we want to output the contents of the object, we first get a description:

```
> trans
transactions in sparse format with
8 transactions (rows) and
6 items (columns)
>
```

Apparently, however, rows and columns exist, so we'll try converting it to a matrix:

```
> as(trans, "matrix")
      apple beer pear chicken milk rice
[1,]  TRUE TRUE FALSE TRUE FALSE TRUE
[2,]  TRUE TRUE FALSE FALSE TRUE
[3,]  TRUE TRUE FALSE FALSE FALSE
[4,]  TRUE FALSE TRUE FALSE FALSE
[5,]  FALSE TRUE FALSE TRUE TRUE TRUE
[6,]  FALSE TRUE FALSE TRUE TRUE
[7,]  FALSE TRUE FALSE TRUE FALSE
[8,]  TRUE FALSE TRUE FALSE FALSE
>
```

For each element, we look to see if it is present or not, and do this for each shopping basket in the rows. We can display the content with the inspect() function:

```
> inspect(trans)
     items
[1] {apple,beer,chicken,rice}
[2] {apple,beer,rice}
[3] {apple,beer}
[4] {apple,pear}
[5] {beer,chicken,milk,rice}
[6] {beer,milk,rice}
[7] {beer,milk} [
[8] {apple,pear}
>
```

The step of transformation into the transactions format is often forgotten; in some cases you also have to consider first which columns of a data frame are suitable at all and how they can be transformed.

9.1.3   Application of the Algorithm

We start the algorithm with the following command:

```
trans.rules <- apriori(trans, parameter=list(supp=0.01,
  conf=0.3,
target="rules"))
```

To the function apriori() we pass the object trans, with additional parameters. These are the minimum values for Support and Confidence which I will cover in the next section. In addition, we specify a target; in this case, rules are to be created. We get the following output:

```
Apriori
Parameter specification:
Algorithmic control:
Absolute minimum support count: 0
set item appearances ...[0 item(s)] done [0.00s].
set transactions ...[6 item(s), 8 transaction(s)] done [0.00s].
sorting and recoding items ... [6 item(s)] done [0.00s].
 creating transaction tree ... done [0.00s].
checking subsets of size 1 2 3 4 done [0.00s].
writing ... [52 rule(s)] done [0.00s].
 creating S4 object ... done [0.00s].
```

With the given parameters, 52 rules were identified. The output is a bit confusing because there are multiple outputs, so to speak, as Figure 9.1 shows. We focus on the first output as printed above.

You can now change the rules with

```
inspect(subset(trans.rules))
```

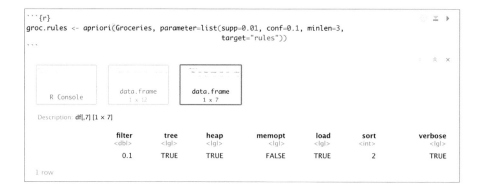

FIGURE 9.1
The output of "apriori" in RStudio.

The result and interpretation will be discussed in the next section.

9.1.4 Interpretations of the Results

In Figure 9.2, each line represents an association rule. The fifth line contains the association of pear and apple. We could express this rule colloquially like this: The occurrence of pear (lhs for left-hand side) when apple (rhs for right-hand side) occurs. A rule, by the way, does not imply causality (although an association of wine and corkscrew might suggest this).

Each rule comes with metrics:

• Support: The number of transactions with this item combination divided by all transactions, regardless of whether other items were in the transaction. The combination milk,beer occurs three times: once alone, twice with other items.

• Confidence: Has nothing to do with confidence, which we have already learned about in the significance test. Here, it is about the support for the simultaneous occurrence of all items in a rule, conditioned only by the support for the left-hand set. This is calculated like this:

$$confidence(X => Y) = \frac{support(X \cap Y)}{support(X)}$$

in the example for

$$apple => pear$$

	lhs		rhs	support	confidence	coverage	lift	count
	<chr>	<chr>	<chr>	<dbl>	<dbl>	<dbl>	<dbl>	<int>
[1]	{}	=>	{milch}	0.375	0.3750000	1.000	1.000000	3
[2]	{}	=>	{apfel}	0.625	0.6250000	1.000	1.000000	5
[3]	{}	=>	{reis}	0.500	0.5000000	1.000	1.000000	4
[4]	{}	=>	{bier}	0.750	0.7500000	1.000	1.000000	6
[5]	{birne}	=>	{apfel}	0.250	1.0000000	0.250	1.600000	2
[6]	{apfel}	=>	{birne}	0.250	0.4000000	0.625	1.600000	2
[7]	{huhn}	=>	{milch}	0.125	0.5000000	0.250	1.333333	1
[8]	{milch}	=>	{huhn}	0.125	0.3333333	0.375	1.333333	1
[9]	{huhn}	=>	{apfel}	0.125	0.5000000	0.250	0.800000	1
[10]	{huhn}	=>	{reis}	0.250	1.0000000	0.250	2.000000	2
1–10 of 52 rows						1 2 3 4 5 6 Next		

FIGURE 9.2
The identified association rules.

this would be

$$\frac{0.25}{0.625} = 0.4$$

This could be expressed like this: Pears appear in 40% of the cases where apples also appear.

- Lift: This metric indicates how much more often a set appears than we would expect if the items were independent. The lift is calculated as follows:

$$lift(X => Y) = \frac{support(X \cap Y)}{support(X) \cdot support(Y)}$$

Illustrated by the example

$$apple => pear$$

this would be

$$\frac{0.25}{0.625 \cdot 0.25} = 1.6$$

The combination apple-pear appears 1.6 times more frequently than we would expect if their occurrence were independent.

These three metrics are important in combination because we want to mine sets of items from the data that come up frequently enough that they provide business value. In addition, you want to see a strong association mapped to Confidence. However, this can also be misleading if we look at the items apple and beer, for example. They are often placed together in a shopping basket, but this may simply be because these items are generally purchased frequently. This effect can be alleviated by the Lift metric.

9.1.5   Visualization of Association Rules

First, let's look only at the rules that have a particularly high lift:

```
> subrules <- head(trans.rules, n = 15, by = "lift")
```

Then we load the package arulesViz which can visualize the data from the rules. We choose the Graph method (I am not mentioning the other methods here):

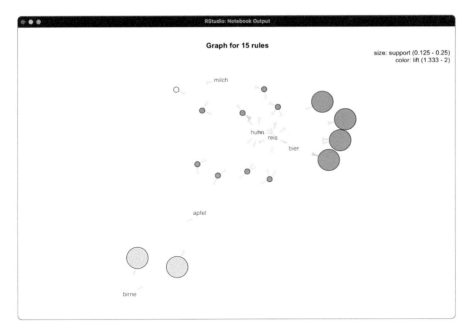

FIGURE 9.3
The visualization of association rules.

```
> library(arulesViz)
Load required package: grid
Registered S3 method overwritten by 'htmlwidgets':
method from
print.htmlwidget tools:rstudio
Registered S3 method overwritten by 'data.table':
method from
print.data.
table
> plot(subrules, method = "graph")
```

You should now see a plot like the one in Figure 9.3.

First of all, we see the individual rules that are made up of items and connections. From an item, an arrow goes to a circle, and from the circle to another item. The darker the circle, the higher the lift. The bigger the circle, the higher the support. The arrows determine the direction, i.e. from left-hand side to right-hand side.

This is only one of several possible visualizations. It is also possible to create an interactive visualization in which the individual elements can be moved using drag and drop. Especially with many rules, the visualization becomes confusing very quickly.

9.1.6   Association Rules with the Groceries Data Set

We now want to apply what we have learned to a larger data set. Nicely enough, the package apriori also comes with a sample dataset, Groceries. This contains almost 10,000 transactions (shopping baskets or receipts) and 169 items (products).

```
data("Groceries")
inspect(head(Groceries))
```

You can see the result in Figure 9.4. Again, we create association rules; it is recommended to experiment a bit with the parameters Support and Confidence to obtain interesting rules in the sense of data mining. At the same time, we use another parameter here, minlen, that defines the minimum number of items in a rule, in this case, three:

```
groc.rules <- apriori(Groceries, parameter=list(supp=0.01,
  conf=0.1,
minlen=3, target="rules"))
Apriori
parameter specification:
Algorithmic control:
Absolute minimum support count: 98
set item appearances ...[0 item(s)] done [0.00s].
set transactions ...[169 item(s), 9835 transaction(s)] done
    [0.00s].
sorting and recoding items ... [88 item(s)] done [0.00s].
 creating transaction tree ... done [0.00s].
checking subsets of size 1 2 3 4 done [0.00s].
writing ... [96 rule(s)] done [0.00s].
 creating S4 object ... done [0.00s].
```

	items <chr>
[1]	{citrus fruit,semi–finished bread,margarine,ready soups}
[2]	{tropical fruit,yogurt,coffee}
[3]	{whole milk}
[4]	{pip fruit,yogurt,cream cheese ,meat spreads}
[5]	{other vegetables,whole milk,condensed milk,long life bakery product}
[6]	{whole milk,butter,yogurt,rice,abrasive cleaner}

6 rows

FIGURE 9.4
Items in the "Groceries" data set.

	lhs <chr>		rhs <chr> <chr>	support <dbl>	confidence <dbl>	coverage <dbl>	lift <dbl>	count <int>
[1]	{curd,yogurt}	=>	{whole milk}	0.01006609	0.5823529	0.01728521	2.279125	99
[2]	{whole milk,curd}	=>	{yogurt}	0.01006609	0.3852140	0.02613116	2.761356	99
[3]	{whole milk,yogurt}	=>	{curd}	0.01006609	0.1796733	0.05602440	3.372304	99
[4]	{pork,other vegetables}	=>	{whole milk}	0.01016777	0.4694836	0.02165735	1.837394	100
[5]	{pork,whole milk}	=>	{other vegetables}	0.01016777	0.4587156	0.02216573	2.370714	100
[6]	{other vegetables,whole milk}	=>	{pork}	0.01016777	0.1358696	0.07483477	2.356750	100
[7]	{other vegetables,butter}	=>	{whole milk}	0.01148958	0.5736041	0.02003050	2.244885	113
[8]	{whole milk,butter}	=>	{other vegetables}	0.01148958	0.4169742	0.02755465	2.154987	113
[9]	{other vegetables,whole milk}	=>	{butter}	0.01148958	0.1535326	0.07483477	2.770630	113
[10]	{other vegetables,domestic eggs}	=>	{whole milk}	0.01230300	0.5525114	0.02226741	2.162336	121

1–10 of 96 rows   1   2   3   4   5   6 .. 10 Next

FIGURE 9.5
Rules created from the "Groceries" data set.

Here, we are at 96 rules; however we are also already at a relatively high Support value given the large number of products. We look at the rules in Figure 9.5.

You may have noticed with this data set that it can take a little longer for the rules to generate, due to the amount of data. Increase or decrease the values for Support and Confidence to get a feel for how this affects the runtime of the algorithm.

What can you do with the results? If you have an e-commerce store, you could suggest suitable products for the item displayed in each case. For a web-site, you could recommend other content. However, the question then arises as to whether, in the long run, you won't simply create trails that reinforce themselves. Besides, the results are based on what the users are actually al-ready doing. Maybe you think your users should do something else and suggest another option?

9.2 k-nearest Neighbors

In the association analysis, we have identified items that are bought together. We now look at an algorithm that does the opposite in the broadest sense, showing us outliers that can also be used for classification.

9.2.1 How the Algorithm Identifies Outliers

We have already learned about the variable k in the section about k-Means. However, k has a slightly different meaning here, because here k does not stand for the number of clusters, but for the number of nearest neighbors (by the way, abbreviated as kNN). This number of neighbors k should help us to identify the similarity or dissimilarity of elements.

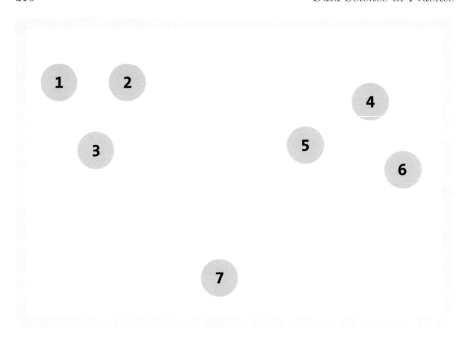

FIGURE 9.6
With k = 2, data point 7 is not a nearest neighbor.

In Figure 9.6 you can see seven data points, and obviously data points 1 to 3 and data points 4 to 6 are closest to each other. Data point 7 is farther away from all the data points. If we choose k = 2, the two nearest neighbors, then data point 7 would still have two nearest neighbors, 3 and 5, but none of the data points would identify data point 7 as a nearest neighbor. Data point 7 is obviously an outlier.

However, if we choose k = 3, then data point 7 would also be a nearest neighbor for more of its neighbors. Data point 5, for example, would have data points 4, 6, and 7 as nearest neighbors. Since this is a very abstract example, it makes sense to take another look at how it works with a more realistic data set. For this purpose, we use the data set with the school children and teachers, here initially only the age, but with a small change: I have made the oldest "school child", who was previously 20 years old, 2 years older, so that this person is now 22 years old.

```
ages <- c(14, 11, 11, 18, 17, 13, 15, 14, 12, 19, 12, 19, 13, 16,
22, 11, 11, 18, 13, 42, 37, 42, 41, 39)
```

We use a library called dbscan, but we don't load it; we just use the function kNN() from it and output the nearest neighbors with k=5. kNN() needs a matrix, so we have to transform our vector:

```
> neighbors <- dbscan::kNN(as.matrix(ages), k=5)
```

The output is a bit longer. First, we are shown the neighbor data points for each school member:

```
> neighbor$id
        1 2 3 4 5
[1,]  8 6 14 20 7
[2,]  3 17 18 11 9
[3,]  2 17 18 11 9
[4,]  19 5 13 12 10
[5,]  19 15 4 7 13
[6,]  14 20 8 1 11
[7,]  15 8 1 5 6
[8,]  1 6 14 20 7
[9,]  11 2 3 17 18
[10,] 13 12 19 4 5
[11,] 9 2 3 17 18
[12,] 13 10 19 4 5
[13,] 12 10 19 4 5
[14,] 6 20 8 1 11
[15,] 7 5 19 4 8
[16,] 13 12 10 19 4
[17,] 2 3 18 11 9
[18,] 2 3 17 11 9
[19,] 4 5 13 12 10
[20,] 6 14 8 1 11
[21,] 23 24 25 22 16
[22,] 25 24 21 23 16
[23,] 21 24 25 22 16
[24,] 21 23 25 22 16
[25,] 22 24 21 23 16
```

Then, the respective distances of the data points to their neighbors are output:

```
> neighbor$dist
        1 2 3 4 5
[1,] 0 1 1 1 1
 [2,] 0 0 0 1 1 [
3,] 0 0 0 1 1 [
4,] 0 1 1 1 1
[5,] 1 1 1 2 2 [
6,] 0 0 1 1 1 [
7,] 1 1 1 2 2
[8,] 0 1 1 1 1
[9,] 0 1 1 1 1
[10,] 0 0 1 1 2 [
```

```
11,] 0 1 1 1 1
[12,] 0 0 1 1 2 [
13,] 0 0 1 1 2 [
14,] 0 0 1 1 1 [
15,] 1 1 2 2 2
[16,] 3 3 3 4 4 [
17,] 0 0 0 1 1 [
18,] 0 0 0 1 1 [
19,] 0 1 1 1 1
[20,] 0 0 1 1 1
[21,] 0 1 3 5 20 [
22,] 2 4 5 5 15 [
23,] 0 1 3 5 20 [
24,] 1 1 2 4 19 [
25,] 2 2 3 3 17
```

We can already make an exciting observation, namely, that for the last five
data points, for the teachers, in each case one neighbor is much further away
than the other neighbors. And as we already know, this is no surprise because
since there are only five teachers, each teacher can only have four neighbors
who are also a teacher – the fifth neighbor must therefore be a schoolchild,
and, as luck would have it, the "schoolchild" who is 22 years old. Also, before,
if the person had still been 20 years old, he would have been the fifth closest
neighbor. If we also look at his row, we see that he also has a higher distance
to the other school children there. Now let's see who has no neighbor here
with a k=3:

```
my_k <- 3
neighbors <- dbscan::kNN(as.matrix(ages), k=my_k) neighbors
<- neighbors$id
neighbors <- unlist(as.list(neighbors))
neighbors <- unique(neighbors[1:(25*my_k)])
length(neighbors)
[1] 24
```

We have 24 elements in the list; however, there are 25 in our ages vector.
One element does not seem to have a neighbor:

```
element <- 1:length(ages)
(outlier <- element[which(!element %in% neighbors)])
[1] 16
```

We had already identified data point 16 as an outlier. Think back again
to hierarchical clustering. There, we had built a distance matrix where the
distance from each point to another point was calculated. Now, we didn't
calculate the distances between each other, but which k neighbors are closest

to each data point. If k = 3, then for each data point, we simply determine the three closest data points, if k = 4, ... well, you understand the principle. Miraculously, we can also immediately apply some of what we learned in hierarchical clustering: If we have variables that differ greatly in their range (for example, age versus height in meters), the data must first be scaled. We did not do this here, because we only used age.

9.2.2   Who Is the Furthest out of Everyone Now?

Let us now think one step further. We can see so far, depending on how we have chosen k, if there are one or more neighbors that fall further out with respect to all neighborhoods. But what if we want to know, summarized in a single number, who are the school members that fall furthest out? That's exactly what the local outlier factor approach does for us. It calculates how far a neighbor is from its neighbors with reference to how far those neighbors are each in turn from their neighbors. Let's look again at Figure 9.6: Data point 7 is closer to data point 5 than to data point 3. Data points 3 and 5 are also farther apart than data point 7 is from either. And even though data point 7 does not have the farthest distance to other data points, this data point is determined to be the outlier. This is because the other data points, while having further distances to each other, have at least locally close distances, unlike data point 7.

Let's apply this to the academic world: If I write a scientific article in information science and am then cited by a scientist who has himself been cited frequently, this benefits my academic career. However, being cited by an eminent scientist is unlikely to get me anywhere for long, because that eminent scientist is likely to cite other eminent scientists much more frequently and to be cited by them as well. I would have some closeness to that one scientist, but not to his community, which cites each other more frequently.

From these considerations, the Local Outlier Factor results in a single number. We use the library Rlof here:

```
library(Rlof)
```

This contains the function lof(), to which we pass the data set, then k, here in a vector, so that I can calculate the Local Outlier Factor for different k, and finally the number of cores I want to use for the calculation, because if we have a lot of data points and a lot of variables, kNN becomes very computationally intensive. The result is a table with the respective data points (school members) in the rows and the LOFs per k in the columns:

```
lof(ages,c(3:8),cores=2)
             3 4 5 6 7 8
[1,]  0.9500000 1.0600000 1.0559896 1.0701823 1.0566277
[2,]  NaN 1.0000000 1.1588542 0.9692708 0.9520833
```

```
[3,]  NaN 1.0000000 1.0000000 1.1588542 0.9692708 0.9520833
[4,]  0.9500000 1.0600000 1.0850000 0.9840686 1.0019225 1.2129188
[5,]  1.1555556 1.4081633 0.9965986 0.9887755 1.0019225 1.1024279
[6,]  1.0000000 0.9444444 0.9444444 0.9031339 0.9686321 0.9815427
[7,]  1.1555556 1.4081633 1.2550097 1.0151561 1.1444446
[8,]  0.9500000 1.0600000 1.0559896 1.0701823 1.0566277
[9,]  Inf 1.0000000 1.0000000 0.9754464 1.0474330 1.0290179
[10,] 1.0000000 0.9166667 0.9422222 1.0957110 1.0719388 1.1091010
[11,] Inf 1.0000000 1.0000000 0.9754464 1.0474330 1.0290179
[12,] 1.0000000 0.9166667 0.9422222 1.0957110 1.0719388 1.1091010
[13,] 1.0000000 0.9166667 0.9422222 1.0957110 1.0719388 1.1091010
[14,] 1.0000000 0.9444444 0.9444444 0.9031339 0.9686321 0.9815427
[15,] 1.5277778 1.5000000 1.3148148 0.9808446 1.1157526 1.1741998
[16,] 3.0000000 3.1733333 2.0305556 1.3821895 1.3360248 1.2960860
[17,] NaN 1.0000000 1.0000000 1.1588542 0.9692708 0.9520833
[18,] NaN 1.0000000 1.0000000 1.1588542 0.9692708 0.9520833
[19,] 0.9500000 1.0600000 1.0850000 0.9840686 1.0019225 1.2129188
[20,] 1.0000000 0.9444444 0.9444444 0.9031339 0.9686321 0.9815427
[21,] 0.9031339 0.9597953 1.8705882 4.0758930 3.6467428 3.0413984
[22,] 1.4779647 0.9597953 1.8705882 3.7095360 3.3178117 2.7652590
[23,] 0.9031339 0.9597953 1.8705882 4.0758930 3.6467428 3.0413984
[24,] 1.0576923 1.0309598 1.8705882 4.0026216 3.5809566 2.9861705
[25,] 1.0713848 1.1021242 1.8705882 3.8560788 3.4493842 2.8757147
```

The result is unsurprising: With k up to and including 4, the school child with the number 16 is the data point with the highest value, followed by the teachers. From k = 5, the teachers become the data points with the highest values, which is not surprising, because now there are at least two school children, so that a higher distance arises. The person with the number 16, on the other hand, no longer stands out as before.

9.2.3   kNN as Classifier

Now that we've looked at how kNN identifies outliers, let's look at how you can use its functionality for classification. Here, the neighbors of a data point determine which class it belongs to. Thinking back to our school example, if k is low, we could use the neighbors to predict relatively accurately whether a data point belongs to a teacher or a schoolchild based on age. However, the larger k becomes, the more school-children become neighbors of a teacher; so to speak, the faster a teacher is also "elected" to be a school child. We can conclude that a class imbalance, as it is present here, is a problem for the algorithm. Therefore, we use a new data set that is about predicting credit risk (Dua and Graff, 2019 [21]). I have pre-processed the data set for you, as the original formatting was a bit awkward; you can find my version on the website for the book. The variables present are:

- result (the target variable, 1 for good risk, 2 for bad risk).

- account balance (categorical)

- duration (of the credit period)

- credit history (categorical)

- purpose (of the credit, categorical)

- credit amount

- savings (categorical)

- busy_since (categorical)

- rate_vs_income (rate as a percentage of disposable income, probably calculated on annual income).

- gender (categorical)

- co-debtor (number of credit applicants for this credit or guarantors)

- registration_address_since (number of years)

- assets (categorical)

- old

- obligations (categorical)

- apartment (categorical)

- existing_credits (number)

- ob (categorical)

- dependent (number of people to whom the credit applicant is dependent)

- telephone (categorical)

- status (categorical)

I did not include another variable, guest workers yes or no, for obvious reasons; it did not matter anyway. In the original data set, status and gender were also combined in one variable, with a distinction between married/widowed or divorced or single for men, and only between single or married/widowed/divorced for women. Apparently, being divorced or not matters more in this data set for men than for women.

We create dummy variables because of our categorical features and then scale the values:

```
library(caret)
result <- credit_data$result credit_data
<- credit_data %>%
select(-result)
dummies <- dummyVars(~ ., data = credit_data, fullRank = T)
credit_dummy.data <- data.frame(predict(dummies,
newdata = credit_data))
credit_dummy.data.scaled <- scale(credit_dummy.data)
Finally, we add the target variables back to the dataset:
credit_data_final <- as.data.frame(cbind(result,
credit_dummy.data.scaled))
```

In the next step, we create training and test data:

```
set.seed(3456)
trainIndex <- createDataPartition(credit_data_final$result,
p = .8,
list = FALSE,
times = 1)
creditTrain <- credit_data_final[ trainIndex,]
creditTest <- credit_data_final[-trainIndex,]
train.labels <- creditTrain[,1]
test.labels <- creditTest[,1] creditTrain <- creditTrain[,
-1]
creditTest <- creditTest[,-1
]
```

For the classification, we use the library CLASS:

```
install.packages("CLASS")
library(class)
```

The function knn for classification is given Probability as an additional parameter, since we want to know what the probability of a class is. We also specify that we want to start with k = 4:

```
knn_results <- knn(train = creditTrain, test = creditTest,
cl = train.labels, prob = T, k = 4)
```

We merge the results with the labels and create a confusion matrix:

```
comp <- cbind(knn_results, test.labels)
 xtab <- table(knn_results, test.labels)
confusionMatrix(xtab)
```

```
   Confusion Matrix and Statistics
test.labels
knn_results 1 2
1 128 52
2 12 8

              Accuracy : 0.68
95% CI : (0.6105, 0.744)
No Information Rate : 0.7
P-Value [Acc > NIR] : 0.7579

                 Kappa : 0.0588

 Mcnemar's Test P-Value : 1.088e-06

           Sensitivity : 0.9143
Specificity : 0.1333
Pos Pred Value : 0.7111
Neg Pred Value : 0.4000
Prevalence : 0.7000
Detection Rate : 0.6400
Detection Prevalence : 0.9000
Balanced Accuracy : 0.5238

        Positive' Class : 1
```

We can see immediately that the results are not good (but we have not optimized a bit here, and there was not a lot of data). We have a very good value for sensitivity, which means that we have detected almost all loans that have been repaid in full. Unfortunately, we have a big problem with specificity: we misidentified almost all the loans that had problems. This would be a disastrous result for a bank, as it would very likely not be able to compensate for the defaulted loans with the fees for the working loans.

9.2.4 LOF for Misclassification Analysis

The question is whether we can use the Local Outlier Factor to see what went wrong here. Again, we use the lof() function from the Rlof library and calculate the LOF for k = 4 and 5, then merge the results with the original data:

```
credit_lof <- lof(knn_results_full[,3:47],k=c(4:5),cores=2)
lof_comp <- cbind(credit_lof,knn_results_full)
```

We display only the first six results:

```
> lof_comp[1:4] %>%
```

```
+ arrange(desc('5')) %>%
+ head()
          4 5 knn_results test.labels
814 2.111263 2.018506 1 2
216 1.906556 1.955661 1 1
204 1.650821 1.745769 1 2
28 1.444221 1.439785 1 1
614 1.403959 1.407212 2 1
434 1.363386 1.387890 2 1
```

Here alone, four out of six lines have been misclassified. Can we see anything here? Data point 814 is a Local Outlier, where it was assumed that the credit risk is low. However, the 814 refers to the row in the original data set; in the test data set, we only have 200 rows. The k-nearest neighbors of data point 814 are:

```
[164,] 131 14 107 113 57
```

The associated distances are (I'll add a few lines before and after so you can classify the result):

```
[162,] 3.339714 4.845647 4.900050 5.539611 5.653733
[163,] 6.336134 6.881541 8.536416 8.774809 8.928773
[164,] 10.193999 10.283622 10.358182 10.495790 10.613166
[165,] 6.190849 6.695545 6.958399 7.369571 7.470682
[166,] 2.679490 4.544855 5.303552 5.556646 5.621172
[167,] 4.466456 6.067669 6.093922 6.113247 6.439973
```

Apparently, this data point 164 is different from its neighbors. If we look at the differences in the characteristics in the data set, we notice the following:

- The customer has specified a household appliance as the purpose, which is rather rare.

- The rate in terms of income is slightly higher than the average.

- The customer was older than average. All other variables seem to be the same. It is not entirely surprising that the algorithm failed here.

We could also use kNN before training and sort out the data that represent outliers. However, you should not take this as a recommendation.

10

Workflows and Tools

This chapter covers everything that has nothing to do with analysis and modelling but is essential in many Data Science projects: from organizing code to analyzing large data sets to deploying interactive applications.

10.1 Versioning with Git

Organizing and storing code is not only important for collaborative teamwork, but also for one's own work. Git has become the most popular tool for software versioning.

10.1.1 Why Versioning?

Imagine that you are working on a new project. You write your code, everything works wonderfully, the code runs through, and you are very happy. You have another idea and try it out, but unfortunately, now the code doesn't work anymore. No problem, you think to yourself. First a coffee to wake up again, then someone else calls and takes you out completely, and when you sit down at the computer again, you no longer know how the old code worked. Of course, you can try to use Ctrl+Z (or whatever undoes what you typed) to restore the old code, but unfortunately, it doesn't work. You are annoyed. If only you had saved your old working code so that you could revert to that earlier version ...

And that's exactly what versioning does. You store versions of your code in a repository, also called a *clone*, so that you can revert to older versions at a later point in time. Of course, this requires you to pause at exactly the right moment and *commit* the current status of your code, as it is called in the world of versioning. After a few weeks, you will have forgotten what you once did in your code, and then it's always good if you can "beam back", so to say, to an earlier version. Of course, you can also solve this by saving different versions of your files. Congratulations if you are well organized, but most of the time, I see file names like "great-code-final-final2.R".

Version management brings even more advantages for the work in a development team. Two people can work on the same code base and merge their results at a later point in time to create a new version.

DOI: 10.1201/9781003426363-10

10.1.2   Git, GitHub, and GitLab

You've probably heard of Linux, the free operating system modeled after the UNIX operating system, which was initiated in the 1990s by the student Linus Torvalds. Linus Torvalds did not simply retire after Linux but invented Git. For the further development of Linux, the previous distributed system for version management of the Linux code could no longer be used, and alternatives existing at that time, such as CVS (Concurrent Versions System), did not come into question. Just like Linux, Git is free software. By the way, Git means something like "dumb ass", and Linus apparently once said that he always names his projects after himself: first Linux, then Git.

Like many Linux tools, Git is a command-line-based tool that is installed on your computer. The complete repository of a project is stored on your computer, even if you only work on a part of it. This is a basic principle of Git, because Git does not need any server structures and instead, uses the distribution on different computers (unless you work alone on a project on your computer). Git even provides its own protocol to synchronize repositories between different machines.

But, this is not the optimal solution for every purpose. For example, if you work alone, what if your computer is broken, and unfortunately, your backup didn't work properly either? Obviously, in such cases, storage on a server can help. And that's exactly what (and much more) is offered by GitHub and GitLab.

GitHub is a commercial service that provides a network-based version management based on Git. Developers can publish their code along with documentation on GitHub and collaborate not only with other developers but also users of a software. Users can report bugs, for example, and other developers can create their own branch, a parallel code base so to speak, develop it further, and later, make it available to the main developer for integration. Thus, GitHub almost becomes a social network for developers. GitHub has been part of the Microsoft Group since mid-2018.

GitLab is a non-commercial service that is installed on your own servers. This is particularly exciting for companies that do not want their code stored on third-party servers. Many features that GitHub offers are also available in GitLab, but the focus of GitLab is more on simplifying DevOps processes.

10.1.3   Basic commands

A common joke among data scientists (and programmers) is that no one really understands Git because everyone only knows the basic commands. In fact, you can get very far with those, and if you rarely use the other commands, you'll forget how they work very quickly. Again, this is not a complete introduction to Git, just the commands you need to survive, but a very good free book on Git in context with R can be found online: it's called "Happy Git with R" [12].

We have already covered a few commands in the first section:

- commit

- pull

- push

- merge

For a better understanding of these commands, we will illustrate the typical flow in working with Git. With

```
git init
```

a new empty repository is created in a workspace, for example the directory of a project. To clone an already existing repository, for example, from another machine, from GitHub, or from GitLab to your hard disk, enter the following command:

```
git clone ADDRESS_OF_REPOSITORY
```

Whether you are starting with an empty repository or have cloned one, if you now add or change files, you will want to combine those changes into a commit at some point. This is done with the command

```
git add
```

Of course, you can also add only part of the files to the repository by specifying the filenames of the files after git add. With the command

```
git status
```

you can now see which files have already been added and which have not. Once you have added all the files you want, you can now start the first commit with the following command:

```
git commit -m "[A smart comment]"
```

We will learn more about the comment later. Now, the commit is in your repository but, it may still need to be synchronized with another repository, for example, on GitLab or GitHub:

```
git push origin main
```

origin is the repository from which you cloned: otherwise, it is the name of the remote server. Additionally, the branch is also specified; in this case, main. A branch is a line of development, which means that two people may be developing two different versions that will be merged at a later time. In this example, however, no new branch is created, but a new version is created in the main branch.

A few more commands will be explained:

- The *fetch* command fetches changes from a remote repository, but in this case, they are not applied to your own code.

- *merge*, on the other hand, synchronizes changes fetched by fetch with its own code. If there are problems, an error message appears.

- The *pull* command performs both fetch and merge.

If errors occur during a merge, the problems mentioned in the error message must first be solved. For example, it may be that two parties have made different changes to the same file. Then, it must be decided which version should be kept, or changes must be made in the files.

10.1.4   Integration with RStudio

The easiest way to work with R and Git or GitHub, respectively, is to integrate Git with RStudio. Here, as a rule, nothing more has to be entered in the command line: RStudio takes care of almost all the details for you. Some configuration work is still necessary, though.

For secure data exchange from your computer to GitHub and back, using so-called SSH keys is is the best method. GitHub eliminated the previous option of using passwords during 2021. SSH stands for Secure Shell, and the SSH key method involves creating two keys, one private and one public, through which one machine can authenticate to the other. First, a key must be created. This can be done in RStudio under TOOLS – GLOBAL OPTIONS – GIT/SVN. Here, you can create an SSH RSA key, if one does not already exist.

In GitHub, click on PROFILE – SETTINGS/SETTINGS and then on SSH AND GPG KEYS (see Figure 10.1). Here, you can paste the SSH key you just created. Now, nothing stands in the way of a secure connection between your computer and GitHub.

After that, first create a repository in GitHub and at the same time, create a README file (see Figure 10.2). You can do this easily by selecting the respective button. You can also choose whether you want your repository to be public or private. At the time of writing this manuscript, keeping a repository private is only possible if you pay a small monthly fee for it, or if you meet other requirements, such as providing open source software. Click CREATE REPOSITORY, and your first code repository is created. How does this repository get onto your hard drive and into RStudio?

In the next step, first copy the address of your repository from GitHub. Attention: Select SSH and not HTTPS here; after all, you just created an SSH key specially (see Figure 10.3).

The easiest way to get the repository into RStudio now is to create a new project in RStudio (discussed in Section 4.2.5). Here, you select VERSION

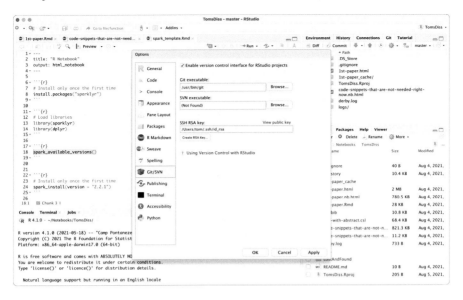

FIGURE 10.1
Creating an SSH key in RStudio.

CONTROL and enter the repository URL exactly as you just copied it from GitHub (see Figure 10.4. RStudio will now do the rest of the work and fetch the files from the repository. Under the hood, RStudio simply executes the necessary Git commands. This is a lot more convenient than typing all the commands yourself.

By the way, you can also choose a different directory than the one suggested by RStudio when cloning; this will then be created as desired. From my own experience, however, it is recommended to choose the same name that the repository has; otherwise, confusion can quickly arise if you have several similar projects on your hard disk.

After cloning the repository in RStudio, you will now see your just-created README in the file browser on the right. If you do not see the README file, something went wrong and you should go through all the steps again.

10.1.5 Commit and Push Code

After you have cloned your first repository, you may want to try the opposite direction. So, not only download code or a README from GitHub, but also put code you wrote locally into your repository – via Commit – and then upload it to GitHub - via Push. I will now describe the steps necessary to do this.

Create a file, be it an R script, an R notebook, or perhaps just a text file, and save it within your project. You should now see a GIT tab in the upper

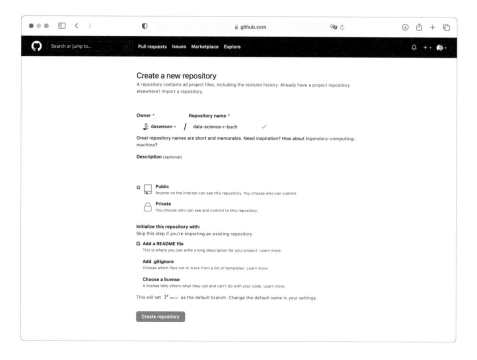

FIGURE 10.2
Create repository on GitHub, including README.

right corner of the workspace area (see Figure 10.5). Click on it so you can see all the files that have been modified since you cloned them. You don't have to remember this yourself; Git will notice all changes and list them for you. However, sometimes, files are created or renewed that you don't necessarily want included. For example, a file called . DS_Store is often found on Macs, and it can quickly sneak into the Git window here as well. Decide which files absolutely belong in the repository; files that you never want to be part of the commits can be added to the .ignore file. Your commits should include code, first and foremost. Larger amounts of data do not belong in a repository.

Important for iCloud users

> If you use iCloud do not put your working directory where the Git repository is located in the iCloud folder. iCloud may have problems with this.

Now, select the files for the commit. This process is called *staging*. It means that you are staging the files from your working directory for the next commit.

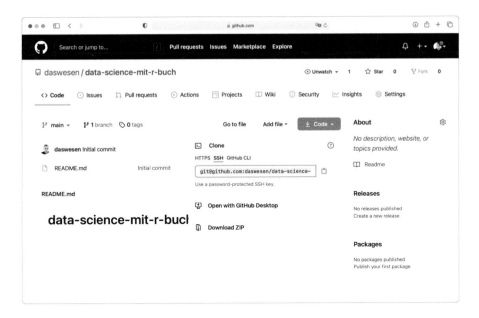

FIGURE 10.3
Copy Web URL of the repository.

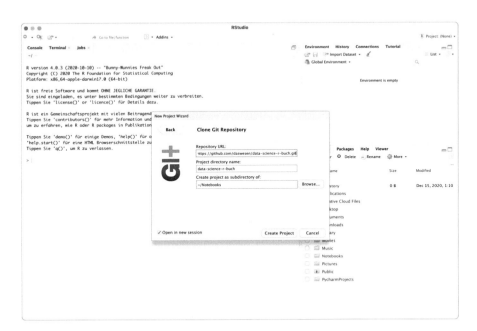

FIGURE 10.4
Create repository in RStudio.

FIGURE 10.5
Selecting the files to be committed under the Git tab in the upper right corner.

You can either select all the files and then deselect the ones you don't want in the commit, or you can simply select the ones that are relevant. Then, click COMMIT.

In fact, you haven't actually executed a commit yet, but you will see a new window where changes can be reviewed. Also, a message about the commit that describes exactly what was changed should be added. This will make it easier for you to revert to exactly the version you need at a point in time. The example in Figure 10.6 is therefore a very good bad example, because this is exactly how you should not do it with this poor comment! If you agree with all the changes, now is the time to click the COMMIT button and thus commit to your repository.

After that, a window opens again showing messages about commit (see Figure 10.7). Normally, you should not see any error messages here. In the figure, Git is complaining because I didn't configure a proper email address. However, you are not finished yet, because with the commit, you have only "pushed" the version of your code into your local repository. Now, you also need to synchronize this version with GitHub as well. This is done with the PUSH button.

Again, you will see a message (Figure 10.8), in this case, that the push was successful. If you see an error here, Google or another search engine you trust will help you. The most common error is that there are differences between the versions, for example, because you have already pushed a newer version from

FIGURE 10.6
A last check before the commit.

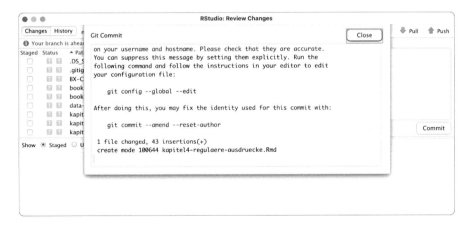

FIGURE 10.7
The "Commit" button is pressed; here it is important that you watch for error messages.

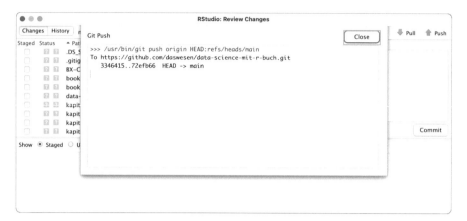

FIGURE 10.8
Message after pushing.

another machine, and GitHub now tells you that you cannot push unless you decide beforehand what to do with the different versions. If you are working alone on one machine, you should never see this error message.

If you use a GitLab-service instead of GitHub, there is in fact hardly any difference. Also, the SSH key is stored in the same way as on GitHub.

10.2   Dealing with Large Amounts of Data

Sometimes, data sets may be bigger than the ones that are discussed in this book. If you get into the gigabyte range, the handling of R with the main memory becomes a challenge. If complex operations are added, every desktop computer is overstrained. There are various approaches to be taken in order to be able to use R even with large data sets of several gigabytes.

One solution may be not to use all the data in a data set, but only some of it. This strategy works well in some cases. For example, a sample from the large data set can be used. However, this does not always make sense.

10.2.1   Need a Bigger Computer? Cloud Computing with R

An alternative to this is offered by cloud servers, for example from Amazon Web Services (AWS). Amazon offers not only one of the largest online shops on the net but also virtual infrastructure. In fact, Amazon is one of the largest providers in the field of cloud computing. Many services on the net are based on AWS infrastructure, be it online storage (Dropbox used to be an AWS customer) or virtual machines running websites. The big advantage is that you only pay for what you need, and only for the time you need it. You can rent a giant server for a few euros for an hour, run large amounts of data on it, and then shut the computer down again so that further use is no longer charged. Setting up such a server is not difficult at all: For the server installation, there are Amazon Machine Images (AMIs) from Louis Aslett for quick installation on Amazon Web Services.[1] Even if you don't need a large server, for a few euros a month, you can thus treat yourself to an RStudio instance that is accessible from any browser. Thus, you can, for example, start an analysis, close the computer, and let the server continue calculating.

Alternatively, you can also use Posit Cloud (more on this in Section 4.2.6). In case of doubt, the Posit Cloud variant is easier to handle, since no operating system updates have to be run. On the other hand, the self-installation on AWS can be cheaper than a Posit Cloud package, depending on how much computing resources is needed. Also, only smaller servers are available in Posit Cloud; AWS offers servers with dozens of CPUs and several hundred gigabytes of RAM. Since R loads all data into memory, a computer with many gigabytes of RAM may result in a faster solution. However, one thing to keep in mind with large amounts of data is that it first has to be transferred to the computer in the cloud. By the way, Kaggle also offers cloud resources but they do not offer RStudio as a cloud service.

10.2.2 Working with Clusters: Spark and Sparklyr

Sometimes, even a larger machine is not enough. In such cases, it may be necessary to parallize processes,, i.e. tasks are divided into smaller tasks and these are processed simultaneously. This immensely reduces the time needed for a task. Apache Spark is a framework for cluster computing that was developed at Berkeley University and has been further developed by the Apache Software Foundation since 2013. On the Spark website, the framework is described as a *unified analysis engine for large-scale data processing*, and especially for large data sets and complex calculations, the linking of R with Spark can provide wonderful support.

A Spark cluster consists of a *manager* and any number of *workers* (see Figure 10.9). Spark takes care of the coordination, so you don't have to worry about it. Tasks can thus be broken down into individual tasks and then distributed among the workers so that they are processed in parallel instead of sequentially. This is obviously not possible for all tasks.

There are two packages to connect R to a Spark cluster: SparkR and Sparklyr. Here, we will focus on Sparklyr because of its proximity to dplyr from the Tidyverse. It requires little additional knowledge to test Sparklyr after reading this section. Sparklyr provides users with an environment that translates almost all dplyr commands for use with Spark into Spark SQL

---

[1] see https://alby.link/louisaslett

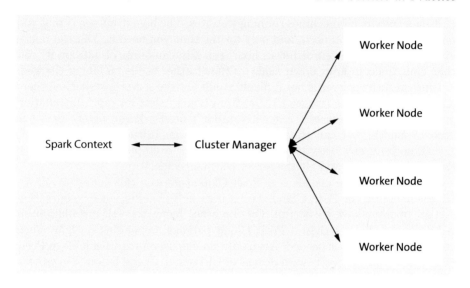

FIGURE 10.9
The structure of a Spark cluster.

statements. At the same time, Spark's machine learning algorithms can be used with Sparklyr.

A Spark cluster can be created locally on your own computer; after all, every processor in modern computers today has multiple cores that can be used in parallel. Many programs use only one core; only a few programs are designed to use more than one core.

You can see how this looks when multiple cores are used in Figure 10.10. Java (the platform in which Spark was developed) uses 576% of the CPU here. Of course, this is not really possible, but since the Mac in the screenshot has eight cores, and I have assigned six cores to Spark, this is how the Mac displays the usage of several cores. The alternative and real purpose of Spark is to use it remotely, on a real cluster with multiple computers. For trying it out for the first time, I recommend the local variant, for which you can also find the code here in the book and on my website.

Pushing your own computer to its limits

Often, the use of Sparklyr leads to the first time you really hear the fan of your computer and realize how hot a computer can get. You can also do without the heater in the room for some projects and heat with your computer alone. Even the new M1 processors from Apple can give off a lot of heat, I've already tried that out for you.

If you are interested in building your own cheap cluster, google "spark cluster raspberry pi." For a few hundred dollars, you can build your own cluster with a few Raspberry Pi-computers to build your own cluster. On YouTube, you will find many experience reports from hobbyists.

As an example of the power of Spark and Sparklyr, the data set nycflights13 is used in many tutorials on the Web, but the use of Spark is not absolutely necessary for this. Any reasonably equipped computer should be able to handle this dataset well, even if it contains more than 300,000 records about flights. However, since not many of you have gigabytes of data available for meaningful data analysis, and I can't make it available for download, this data set will also serve as an example in this book.

First, install the package Sparklyr, which you then load at the beginning of the script as well as dplyr:

```
library(sparklyr)
library(dplyr)
```

It may seem strange that dplyr also needs to be loaded, but sparklyr is actually just a translator of the dplyr commands into the Spark SQL statements. If you don't already have Spark installed, you now need to install it.

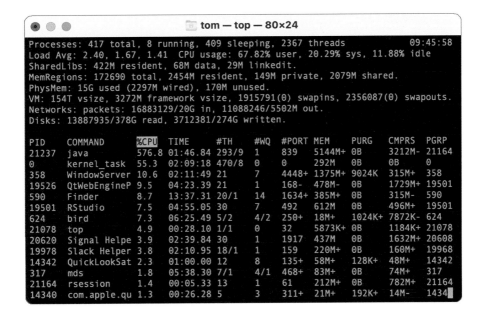

FIGURE 10.10
Six of eight cores are used by Spark.

This first requires a Java Runtime Environment, which you may need to install, ideally before Sparklyr. Java versions for your operating system can be found at https://java.com. First find the current version of Spark; you can find it with the command

```
spark_available_versions()
```

To install Spark, use the following command, adjusting the version number to the current version:

```
spark_install(version = "3.1")
```

Next, define a configuration. Note that you should provide Spark with slightly less RAM than is available in the respective computer so that there is still some RAM left for the operating system. The operating system takes what it needs anyway, but a computer won't get faster if multiple applications are competing for memory.

In the following example, six cores are also used, although my computer has ten cores as shown in Figure 10.11 has 10 cores as shown in Figure 10.11. The other cores are reserved for the operating system and other applications. In addition, R and RStudio require processor time. Please check in advance

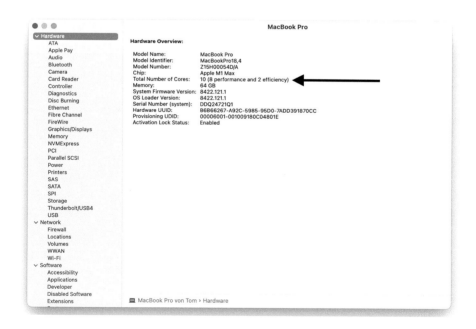

FIGURE 10.11
The number of cores is displayed somewhat hidden under MacOS.

how many cores your computer has! On a Mac, this can be done, for example, via the Apple menu: ABOUT THIS MAC and SYSTEM REPORT. There, under HARDWARE, the number of cores is listed.

```
conf <- spark_config()
conf$'sparklyr.cores.local' <- 6 # number of cores
conf$'sparklyr.shell.driver-memory' <- "14G"
conf$spark.memory.fraction <- 0.9
```

Now, a connection can be established:

```
sc <- spark_connect(master = "local",
version = "3.1",
config = conf)
```

The sc (for spark cluster) object created here is the connection to the Spark cluster. In any subsequent command that you want to do something with the Spark cluster, you must reference this object. You don't have to name the object sc, but the following steps will be a bit more error-prone if you decide to use a different name.

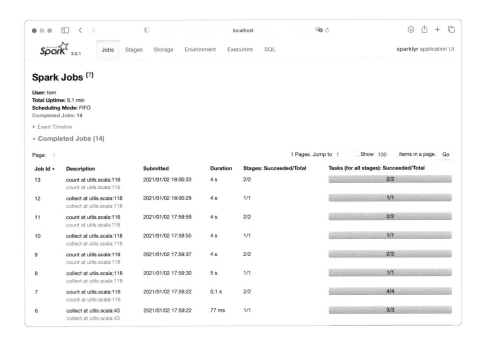

FIGURE 10.12
The Spark Web interface.

Spark provides a web interface after startup that can be accessed at the local URL http://localhost:4040/jobs/ (see also Figure 10.12. This interface is nice if you want to check where your cluster is or what jobs it is currently processing.

This overview is also helpful if you want to understand how Spark handles your memory. In particular, if you use very large data sets, Spark will also swap to disk.

You may still have the flights data set installed from Chapter 5, "Exploratory Data Analysis", if not, please install it now:

```
install.packages("nycflights13")
```

You can load the associated library, but with the following command, it is not necessary:

```
flights_tbl <- copy_to(sc, nycflights13::flights, "flights")
```

Let's take a closer look at this line. We create a new object flights_tbl, which is initially just a list. This list contains references to the Spark data frame "flights" that we copied from the original data frame nycflights13::flights into the Spark cluster via the connection sc. In the environment, you see FLIGHTS_TBL; under the CONNECTIONS tab, you see the Spark data frame FLIGHTS (see Figure 10.13). This may be confusing at first, hopefully it will become a bit clearer once you enter the two objects into the console, first flights:

```
> flights_tbl
# Source: spark<flights> [?? x 19]
```

I'll refrain from outputting the whole table here; instead, I want to draw your attention to the first row of the output. Because what you see is that with this command, you have actually output the Spark data frame flights. And now we're just going to enter flights:

```
> flights
Error: object 'flights' not found
```

We can only access the Spark data frame through our just-created object. This is because the big difference between this and our normal data frames is that the data in our Spark cluster is not loaded into the memory that R uses. If you load a 2-GB file into R, you will see this in R's memory usage. If you load such a large file into your Spark cluster, you do not load it into R's working memory. R then only gets the data on demand via a Spark command, but it always stays in Spark and is not accessible within R itself.

You can also view the table of the Spark data frame as usual by clicking on the table icon, but here not in the Environment but under CONNECTIONS. The icon is again on the far right. However, you will see a maximum of 1,000 rows here, unlike a data set that you have loaded in R. Now, let's work with the data. We will use an example from the Sparklyr documentation:

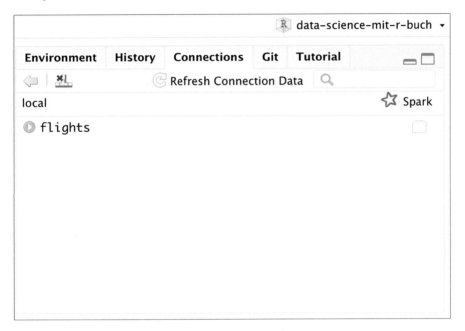

FIGURE 10.13
The Spark data frame "flights" in the "Connections" tab.

```
flights_delay <- flights_tbl %>%
    group_by(tailnum) %>%
    summarise(count = n(), dist = mean(distance),
    delay = mean(arr_delay)) %>%
    filter(count > 20, dist < 2000, !is.na(delay))
```

Nothing special so far; you may be a little surprised at how quickly this command was executed. In fact, it was not really executed at all. Because if you now enter

```
flights_delay
```

then the computational work begins, and the answer is likely to take a little longer. This approach is called lazy loading, and you can see the result in Figure 10.14. In the upper left corner it says LAZY QUERY, and instead of the number of rows, you see only a ??, whereas the number of columns is specified exactly. Spark doesn't know at this point how many rows there would be once everything is calculated through.

This is something you have to get used to, because if any calculations are being done all the time (or at least, you think they are), but then you actually want the data output and have to wait a long time, the machine behaves differently than expected. However, lazy loading has many advantages for

Source: **lazy query [?? x 4]** Database: **spark_connection**			
tailnum <chr>	**count** <dbl>	**dist** <dbl>	**delay** <dbl>
N24211	130	1330.2615	7.700000000
N793JB	283	1528.5972	4.722419929
N657JB	285	1285.8596	5.031802120
N633AA	24	1587.0417	-0.625000000
N9EAMQ	248	674.6653	9.235294118
N3GKAA	77	1246.5974	4.973333333
N997DL	63	867.7619	4.903225806
N318NB	202	814.3465	-1.120000000
N651JB	261	1407.5287	7.580769231
N841UA	96	1208.4271	2.104166667

1–10 of 1,000 rows   1   2   3   4   5   6 ... 100 Next

FIGURE 10.14
The output of the query to the Spark cluster; please note the "lazy query" is
striking here.

efficiency reasons, because at the time of the calculation, only the data that
is actually needed is fetched. And at the first output, flights_delay does not
use any memory at all! By the way, also, the output of flights_delay does not
use any memory of R; only Spark needs it.

Now, let's plot the data once:

```
flights_delay %>%
    ggplot(., aes(dist, delay)) +
    geom_point(aes(size = count), alpha = 1/2) +
    geom_smooth() +
    scale_size_area(max_size = 2)
```

Unfortunately, we now receive an error message:

```
Error: Mapping should be created with 'aes()' or 'aes_()'.
Run 'rlang::last_error()' to see where the error occurred.
```

The reason is that ggplot expects an R data frame or a tibble, but instead
gets a list, an object that references the Spark database result. What do we
need to do to get the result in R?

```
delay <- flights_tbl %>%
    group_by(tailnum) %>%
    summarise(count = n(), dist = mean(distance),
        delay = mean(arr_delay)) %>%
    filter(count > 20, dist < 2000, !is.na(delay)) %>%
collect()
```

Here, everything should already be obvious, except for the last line. With "collect", the result of our expression is copied into the R memory and here, assigned to the object delay. Only then can the graphic be created from the results.

---

collect or collect()?

You can write collect() without parentheses in this case, but somehow it doesn't feel right, since it's a function, and we don't want to get into the habit of doing something like that, even if it's allowed in this case.

By the way, be careful with some functions that work with dplyr but not in combination with sparklyr. For example, unique() outputs something different than you would expect, and none of the stringr functions work either. For unique(), you can use distinct(), and for regular expressions you can use the regexp family, for example, regexp_replace(). The syntax is a little different. See the notebook for this chapter for examples.

We've only touched the tip of the iceberg here. But should you ever come across a larger amount of data, this will at least give you an entry point into the world of cloud computing.

10.2.3 data.table

I have already introduced a powerful extension to R in the form of the Tidyverse (Chapter 5, "Exploratory Data Analysis"). However, this is not the only extension to R, and data.table is seen as a counterweight to the Tidyverse by some who see dplyr as "overhyped". data.table was started by Matt Dowle and has its own website at https://rdatatable.gitlab.io/data.table.

Even if the respective supporters are now arguing about which extension is better, Tidyverse or data.table, you can simply start the experiment yourself once and check which is faster for you; because data.table has two goals: on the one hand, to reduce the time of programming, and on the other hand, to minimize the time for calculation. In the repository for this chapter, you will also find some examples of data.table.

10.3 Deploy Applications via API

If the result of a Data Science project is to be integrated into other software, an API is usually required. The acronym API stands for application programming interface, in contrast to graphical user interface (GUI). The difference is quite simple: You, as a user, usually interact with a computer program with a GUI; you click on icons or type something into a text editor. If a computer program wants to interact with another program, it lacks the means to make use of the GUI. To be able to communicate from machine to machine or program to program, an interface is needed, and that is the so-called API.

An example that is helpful to understand: Not every news website that includes a weather forecast has its own weather stations and meteorologists. The data is pulled from weather data providers. Now, instead of an intern typing the weather report from these providers, the weather data can be automatically retrieved from the weather service by the news website via an API.

10.3.1 What Is a REST API?

A REST API is a popular form of interface that uses common Web development methods to make resources available to another program. In R, models or results of models can be made available via a REST API with a few tools. Such an API or a RESTful API – I will come to the difference in a moment – follows the principles of REpresentational State Transfer and thus, the paradigms of the World Wide Web. A REST API does not provide a protocol of its own and is in fact not a standard of its own, but it uses known standards like HTTP or JSON, which is why such APIs are called RESTful. This includes the fact that every time a client makes a request (the program making the request), all the information needed by the server is present, since the server cannot access other information that was previously sent. Many services on the web work this way, for example, by appending parameters to a URL. An example is the SISTRIX API, which asks for the API key (the password to use the API, so to speak) every time it is called:

```
https://api.sistrix.com/credits?api_key=[API_KEY]
```

REST APIs have other principles, but they are not further relevant for our little experiments here.

10.3.2 Provide an API with the "plumber" Package

The currently most convenient way to provide an API with R utilizes the package "plumber". In the following code snippet, the necessary information for the API is recorded in comments. We will first look at a simple example from the package's vignette. First of all, we create a file with the code for the different endpoints of the API:

```
# plumber.R
#* Echo back the input
```

```
#* @param msg The message to echo
#* @get /echo
function(msg="") {
list(msg = paste0("The message is: '", msg, "'"))
 }  #*
Plot a histogram
#* @serializer png
#* @get /plot
function() {
rand <- rnorm(100)
hist(rand)
}
#* Return the sum of two numbers #*
@param a The first number to add
#* @param b The second number to add
#* @post /sum
function(a, b) {
as.numeric(a) + as.numeric(b)
}
```

The API provides three endpoints here – echo, plot and sum – which are functions, respectively. The exciting part comes in the lines before them, which start with #*. Let's take a closer look at the last example.

Two parameters are determined, a and b. In the third line, a request method of the HTTP protocol is used. The two most common request methods are POST and GET. GET is used to get content from websites whereas POST is used to change data on the server, for example by submitting a form. In fact, the GET request method could have been chosen in the third example, but presumably, the developer wanted to demonstrate that the API could handle more than just GET.

Save the code just shown in an R script (which you can find on my website, of course), ideally named plumber.R. Do not use a notebook for this purpose. Now enter the following command in the console:

```
> library(plumber)
> pr("plumber.R") %>%
+ pr_run(port=8000)
```

You will see the following answers:

```
Running plumber API at http://127.0.0.1:8000
Running swagger Docs at http://127.0.0.1:8000/__docs__/
```

Also, a new window should have opened that looks like the one shown in Figure 10.15. This is a Swagger-service, a collection of tools for creating

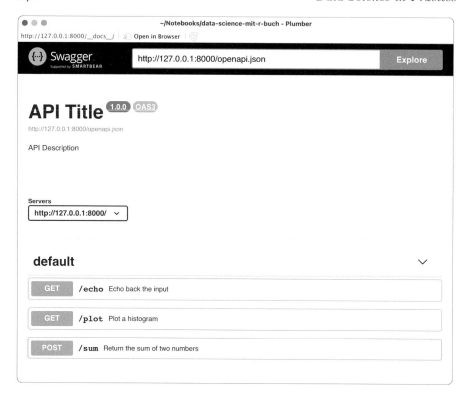

FIGURE 10.15
The Swagger window for testing the API.

and documenting Web services like our API. The great thing is that you can already test your API here.

Of course, you can also test the API in the browser. For example, enter

`http://127.0.0.1:8000/echo?msg=Hello`

in the browser. You will then see the API response in the browser window (Figure 10.16). This is in JSON-format. However, enter

`http://127.0.0.1:8000/echo?sum?a=5&b=6=Hello`

and then, you will most likely get an error message. This is because the POST-request method is used. Change the code to GET, and then, this example will also work properly.

So far, the API lives only on your computer. This is, of course, impractical, because if another program wants to access your API somewhere in the network, it will most likely be difficult (you can, of course, share the port on your router, but I strongly advise against that for security reasons).

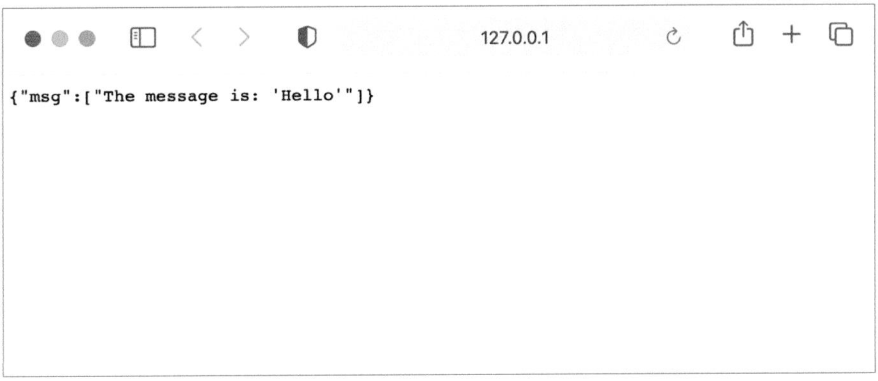

```
{"msg":["The message is: 'Hello'"]}
```

FIGURE 10.16
The response of the API with your endpoint "msg.".

You can publish your API on RStudio Connect, but this involves an expensive paid service. Alternatively, for a few dollars per month, you can try a service like DigitalOcean. For this, the package plumber offers a companion package called plumberDeploy.

10.3.3 The Next Step: Docker

Anyone who develops software knows this problem well: You develop a great program, and it runs fantastically on your own computer. Then, it's released, and someone else gives it a try, comes back to you, and says it doesn't run on their computer. "Works for me" is a popular response, but probably not the most helpful one. Other systems often have a different configuration than the development system, which prevents software from working on that system.

This problem is solved, at least in the context of server applications, with Docker. Here, applications are deployed in containers that contain all the necessary packages. The so-called runtime environment is provided in the container, and the application with its environment is virtualized. You can imagine this roughly as if Windows 10 were running on a Mac. In fact, it works with additional software. Windows 10 then runs in a virtual machine that runs on the Mac. The virtual machine provides everything for the Windows operating system to think it is installed on a PC. This is exactly how a Docker container works: Everything is there for an application to have what it needs to function.

An image for creating a container for R and plumber are available at https://hub.docker.com/r/rstudio/plumber. You can use this to provide a container that contains your API. For more information about Docker, visit docker.com.

10.4   Create Applications with Shiny

The result of a model or an analysis can be provided as a notebook, pure code in a repository, or via an API, as already shown. Another option is to make it available in an interactive web application.

10.4.1   What Is Shiny?

Shiny, like RStudio, is a development from RStudio PBC and is both a package and a server environment. A Shiny server offers the possibility to provide dashboards and models for interaction for users on a web server, so that they are accessible under one URL. The effort needed for an application is low; a dashboard or an application is created with only a few lines of code.

As an example, we will use a Shiny application that comes with RStudio, so to speak. To do this, simply select FILE – NEW FILE – SHINY WEB APP from the RStudio menu bar. The editor will then open the code of an app that contains a simple interface for manipulating a histogram of the Old Faithful Geyser data (see Figure 10.17).

FIGURE 10.17
RStudio changes when a Shiny app is loaded into the code editor.

What is the Old Faithful geyser dataset about?

This geyser is located in Yellowstone National Park in the United States of America. The dataset contains 272 observations of an eruption of the geyser as well as the length of the eruption and the waiting time for an eruption, both in minutes. By the way, it is not necessarily clear in which direction to think here: Is it the waiting time before or after an eruption?

Of course, a Shiny app can also be written to an empty R script file. The extension for a script as well as for a script is .R, because a Shiny app is nothing else than an R script. However, the template is convenient, because the most important elements of a Shiny app are already present.

No matter which Shiny app you open in the editor, it changes the RUN in the interface to RUN APP (see Figure 10.18). If you click on it, the app will be launched from the editor unless an error occurs. A new window opens showing the running app (Figure 10.18).

Play with the slider a bit to get an idea of how the interactive handling of the small data set feels.

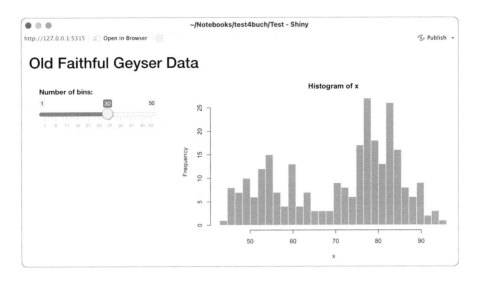

FIGURE 10.18
The first Shiny app.

What should the diagram tell me?

In case you were wondering what you see in Figure 10.18 exactly: A histogram is a graphical representation of a frequency table. That is, you see on the x-axis the waiting time for the outbreak and on the y-axis the frequency of the respective waiting times. The big question with a histogram is how coarse or granular the division of the values should be. The number of "bins" is one such division. You can learn more about histograms in Chapter 5, "Exploratory Data Analysis."

Under certain circumstances, it can happen that a gray veil covers a Shiny app and a message is displayed at the bottom left: DISCONNECTED FROM THE SERVER (see Figure 10.19). This error message can have several reasons:

- A timeout has occurred, which is defined in the application settings.

- Objects that are needed for the app to run are missing.

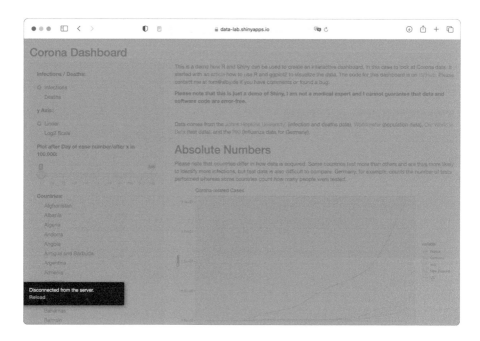

FIGURE 10.19
Not a bug, but a feature: "Disconnected from the server."

- It takes too long to load something, for example, data.

- Another, rather rare error has occurred.

The timeout also happens when you haven't done anything, or just because. After 10 minutes or whatever is defined, this message appears automatically because no more user input was registered. As they say, "It's not a bug, it's a feature."

10.4.2   UI and Server

We'll now take a closer look at the code for this simple example. A Shiny app consists of two components, a user interface (UI) and a server. This is an important concept because there is interaction between the two components. The UI component takes input, passes it to the server, and then presents the server's output back to the interface. The server takes the input, performs operations, and passes the result back to the UI.

First, the library shiny is loaded:

```
library(shiny)
```

Now, the user interface is defined:

```
ui <- fluidPage(
```

The title defined here can be seen later in the browser tab:

```
title titlePanel("Old Faithful Geyser Data"),
```

For the left side, a sidebar is defined, in which a slider is defined. The slider has the name "bins", and the name sliderInput() is predefined.

```
sidebarLayout(
sidebarPanel(
sliderInput("bins",
"Number of bins:",
min = 1,
max = 50,
value = 30)
),
```

The sliderInput function also defines where the slider starts (min), where it ends (max), and what the start value is (value). The main panel is defined for the right side of the browser window. There is a plot output, here again with a predefined function, plotOutput(), whose output has the name distPlot:

```
mainPanel(
plotOutput("distPlot"
) )
    )
 )
```

From here on, the server logic is defined. The server is a function that uses a value as input and outputs values:

```
server <- function(input, output) {
```

And now, it gets exciting: The input we had received via the sliderInput() function in the UI becomes the input for the server here. input$bins contains the input from the UI and is used to create the histogram. The output of the server is defined as a new object, output$distPlot, which – surprise – was already defined in the UI:

```
    output$distPlot <- renderPlot({
# generate bins based on input$bins from ui.R
        x <- faithful[, 2]
bins <- seq(min(x), max(x), length.out = input$bins + 1)
# draw the histogram with the specified number of bins
hist(x, breaks = bins, col = 'darkgray', border = 'white')
})
}
```

Finally, the shinyApp() function creates a Shiny app object from a UI server pair:

```
shinyApp(ui = ui, server = server)
```

Ready is the first Shiny app! Click RUN APP in RStudio interface, and you can admire your first app. Of course, your own share is still small here, and usually, you may want to display something other than a histogram. You can find inspiration in the demo applications of Shiny, which you can start after loading the library:

```
runExample("01_hello") runExample("
02_text") runExample("
03_reactivity") runExample("
04_mpg") runExample("
05_sliders") runExample("
06_tabsets") runExample("
07_widgets") runExample("
08_html") runExample("
09_upload") runExample("
10_download")
```

If you're up for a task: Why not build a Shiny app where users can enter age, gender, and passenger class, and your app will tell you whether the user would have survived on the Titanic (see Chapter 8, "Classification"). To do this, you need to save your pre-created model and then load it into the app. You can see exactly how this works in the notebooks accompanying this book at https://alby.link/datascience. Small tip when creating Shiny apps: Pay attention to the output in the console. If anything goes wrong, you'll see it here - but only for local operation.

10.4.3 Publish a Shiny App from RStudio

If the Shiny app runs on the desktop, it cannot be made available to other users. We already know this problem from the API. A server is required to make the Shiny app available. There are several options for this.

The simplest option is that a Shiny server is already available. In this case, it is enough to log into RStudio, create a new directory ShinyApps in your own directory, and copy the app into it. The app can then be found under <servername>/shiny/<username>. It is recommended to test the app on the server from your own user directory first, but additionally from the URL, where the app is then available to everyone. It is not uncommon, after all, for an app to only work when the user is running it from within RStudio. This is because the app under the URL is not executed by the RStudio user, and thus, not all resources may be available. You need to pay particular attention to absolute paths here. It's worth looking at the log files, because if you call the app from the URL, you won't see any output in the console. You don't even see what went wrong in the first place. The log files are in different places depending on the server; I would first look under /var/log/ on UNIX systems.

Another option is to use one of Posit's products: one is Posit Connect, the other shinyapps.io. I explain the latter in Section 10.4.5. This might be the easiest way to test and deploy a Shiny app on a server.

10.4.4 Example Applications

In March 2020, Corona had finally arrived in Germany. At that time, many different numbers were thrown around, each with their own interpretations. Since the data was generally publicly available, it was easy to create a custom Corona dashboard using Shiny. First, we need to load all the data, which was made available by Johns Hopkins University in its own GitHub Repository. In a second step, the data needs to be transformed before the actual Shiny app is is launched.

Using resources sparingly

The method just described is not a best practice, because the data is then really loaded every time the dashboard is called, which puts unnecessary load on the servers involved. However, since the dashboard was just a finger exercise for which not many users were expected, I made no further effort to optimize this. In an ideal world, the data would only be downloaded once a day, so the script would have to look to see if the current data had already been downloaded to its own server, and only load it from the GitHub repository if it hadn't. But since the data is on GitHub in both cases, this should not be such a big problem.

The result can be seen in Figure 10.20 as well as in the link https://alby.link/coronashiny. The code is on GitHub, you can find it via the short link https://alby.link/coronashinygithub.

10.4.5 shinyapps.io

You may have noticed in the previous section that the Corona dashboard is not hosted on my website, but on shinyapps.io. This is another offering from Posit PBC, the company that also developed RStudio and Posit Cloud. Instead of having to install and maintain your own server, everything is already taken

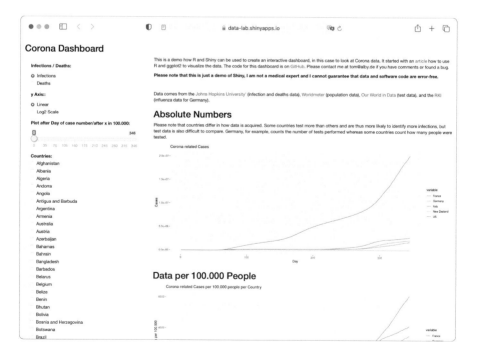

FIGURE 10.20
The Corona Dashboard.

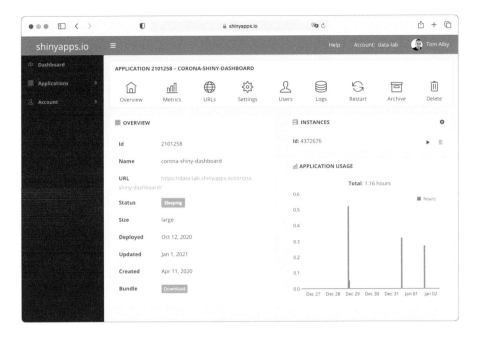

FIGURE 10.21
The administration interface of shinyapps.io.

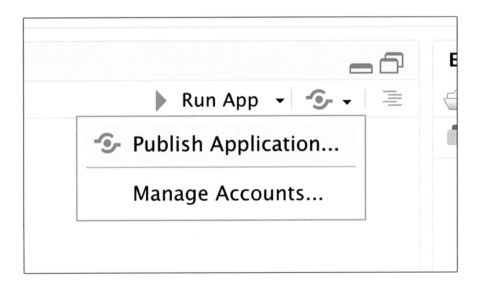

FIGURE 10.22
In the RStudio interface next to "Run App" is a somewhat inconspicuous button for publishing a Shiny app.

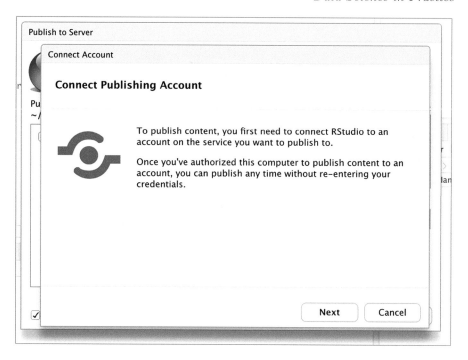

FIGURE 10.23
If no account has been set up yet, this step is inserted now.

care of here. At the time of manuscript creation, a free package exists that can accommodate five apps. This option is recommended because it allows you to follow all the steps described here in the book beyond just creating the app.

In larger, paid packages, it is possible, for example, to use a separate domain or to require user authentication for access. Also, here ,you can choose different sizes of the instance on which the Shiny application will run. In addition, you can modify the timeout limit.

All these settings would have to be made laboriously in the server settings or when choosing the instance in the case of your own installation. On shinyapps.io, you can even switch from one instance size to another relatively easily.

However, the free package already offers some great options to create a professional-looking app. Debugging options also exist. You can view the log files of each app via the Web interface of shinyapps.io and thus, check relatively easily what might have gone wrong if your app should ever fail to run properly.

The process of publishing an app from RStudio is almost automatic. You just set up an account with the target service, and then, you can publish your app with the click of a button.

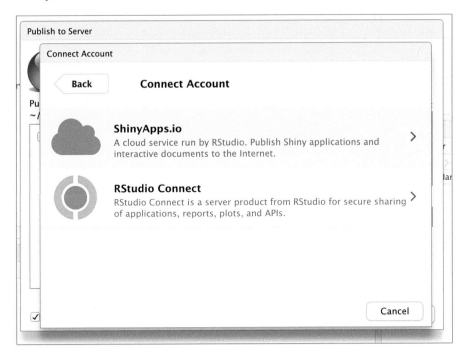

FIGURE 10.24
Choice between ShinyApps.io and RStudio Connect.

If no service or account has been set up yet, this step will be inserted automatically once you have clicked the button to publish an app (Figure 10.24).

You can set up more than one account in RStudio, which is very useful if, for example, an app is to be set up on a staging server first. Such a staging

FIGURE 10.25
The token, here with the invisible Secret.

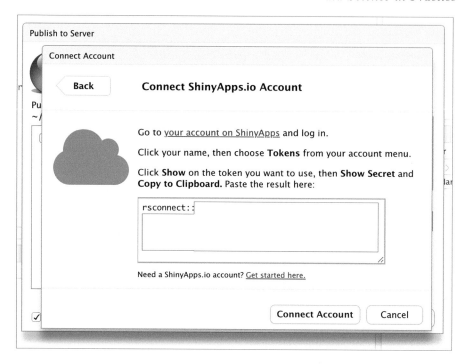

FIGURE 10.26

The token has been removed from the screenshot.

server is used to test software in an environment that resembles the actual live environment. Likewise, you can use several different accounts for different projects.

In order for RStudio to connect to shinyapps.io, a token is required. This can be found in the shinyapps.io interface under ACCOUNT – TOKENS. You can create multiple tokens. A token contains a "Secret", which you can only see if you click on SHOW and then SHOW SECRET (see Figure 10.25)!

You have to copy the whole text in the window you see now and paste it later into RStudio (see Figure 10.26, in this example, I removed my token of course).

Click CONNECT ACCOUNT so that it can be verified that the connection is working.

Last but not least, give the app a name and choose which of the files to publish. If you then press the PUBLISH BUTTON, it may take a few minutes before the app can be accessed on shinyapps.io under its own URL. Even on these last meters, problems can still occur; for example, an image may not run on an instance. In this case, it is worth taking a look at the logs, as described earlier, to find out what the reason for this might be.

If everything is running, then you may wonder why it takes so long for your app to be visible when you call the URL later. This is because your app

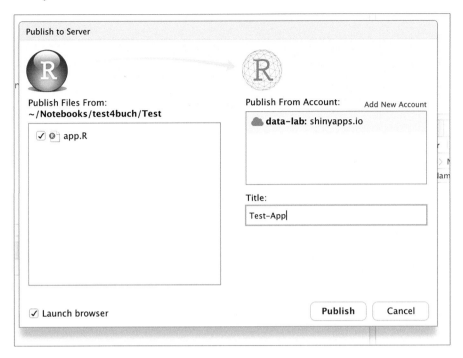

FIGURE 10.27
Selection of the files that will be selected and a name for the app.

is not running all the time, but is "put to sleep" when not in use. You can see this in the interface of shinyapps.io: apps are either "running", "sleeping," or "archived". Depending on which package you took, your app may take a little longer to wake up, with the free package taking the longest. The time your apps are "running" is included in the calculation of "Active Hours". Each package offers only a certain number of "Active Hours". For the first experiments, however, the 25 hours of the free package should be sufficient in any case.

11

Ethical Handling of Data and Algorithms

11.1 Privacy

As Data Scientists, we are more likely to have an understanding of what happens to data than the average citizen. This also brings with it a responsibility, because no matter what kind of data you want to process, just because the data is available to you, or because something works technically, this doesn't mean that it was also legally allowed to collect that data, let alone use it at all. No data scientist wants his employer, a client or himself to be sued because user data can suddenly be downloaded from the web or because it turns out that data was used that should not have been collected at all in the first place. Also, the idea that a designed model could discriminate against people, as in the example mentioned, will hopefully make all data scientists aware of the responsibility that rests on them.

Please do not skip this box

Privacy legislations differ from country to country, and in this section, we will review a few concepts that are shared around the world. Having said that, this is not legal advice.

Any data protection issue that arises in a project must be consulted with a knowledgeable professional.

11.1.1 Legislations around the World

While no global privacy legislation exists, data-driven projects that are used in more than one country will either have to adjust algorithms based on a region's legislation, or design data usage in a way that is compatible with all regions according to the lowest common denominator.

The European Union *General Data Protection Regulation (GDPR)* is probably the most famous privacy legislation in the world. It was published in 2016 and came into force after a transition period in 2018. The regulation was adapated by each EU country, based on the following principles:

- Lawfulness, fairness and transparency: Data subjects whose data is to be collected need to be asked for permission, and the processing of the data needs to be explained in an understandable language. Obviously, this raises questions how the understandability can be measured, and the processing within a machine learning model may be difficult to explain. At the same time, data that was collected at an earlier point in time may not be used without a GDPR compliant consent.

- Purpose limitation: If permission to use data has been granted for a specific purpose, the data may not be used for any other purpose. Example: An online retailer obviously needs to process data about products sold and their buyers in order to be able to issue an invoice and, if necessary, also to process warranty claims at a later date. If permission has not been granted for further purposes such as processing the data to improve recommendations on the website (or somewhat more superficially "to improve the offer"), the data may not be used to begin with. And this also applies if the data is used for a shopping cart analysis, from which no offers are to be generated for the data subject at all, but only to improve the quality of recommendations for other customers. The question is whether the use is again permitted if the data is completely anonymized. How can it be ensured that no recommendation can be traced back to an individual user? The example of the AOL data set, which I briefly described in one of the previous chapters, shows that this is not so simple.

- Data minimisation: This means that, as a rule, it is not legally permitted to collect all the data that you find interesting. You must therefore consider very carefully in advance why you are collecting which data. Data that was previously collected without fulfilling this condition must not be used.

- Accuracy: Data must be accurate, and consumers have the right to ask for their data to be corrected if it is wrong.

- Storage limitation: In the example of the retailer, the data on the person can therefore be stored for as long as is necessary, for example, to carry out an exchange under the statutory warranty, provided no other consent has been given. Whether the data can then be used anonymously in a Data Science project, however, must be answered by a data protection officer.

- Integrity and confidentiality (security): In the next sections, we will discuss a few examples where data was "leaked". Data scientists are not necessarily security experts at the same time, and if data and models are located in a cloud, experts must also check whether it is secured in such a way that it cannot be accessed by external parties.

- Accountability: The data processor is fully responsible to comply with the law and needs to be able to demonstrate this compliance.

In addition, there are personal data requiring special protection, also called "special categories" of data:

Processing of personal data revealing racial or ethnic origin, political opinions, religious or philosophical beliefs, or trade union membership, and the processing of genetic data, biometric data for the purpose of uniquely identifying a natural person, data concerning health or data concerning a natural person's sex life or sexual orientation shall be prohibited.

Here, too, there are exceptions, but within very narrow limits.

The *California Consumer Privacy Act (CCPA)* came into effect in 2020 and can be regarded as a counterpart to the GDPR. Having said that, as the name says, the CCPA only affects California and not the rest of the US. However, several data-consuming companies such as Google, Facebook, and Apple have their headquarters in California. In contrast to the GDPR, it is limited to companies with a minimum of turnover or a ratio of revenue made by selling data or the collection of data of a threshold of households in California.

The law is, similar to the GDPR, based on a few basic principles:

- Right to Notice: Similar to the GDPR, users need to be informed about a company's practice of collecting, using and sharing personal information when the data is collected.

- Right to Erasure: Organizations are required to delete all data if a consumer requests it; this needs to happen within 45 days.

- Right to Opt-in for Minors: Data from minors is not to be sold unless the minor or a parent allows it.

- Right to Continued Protection: Even if data is sold, consumers have the right to have their data deleted by the party that has acquired the data.

- Right to Awareness: Businesses must specify exactly the consumers' right to have their data erased, collected or sold.

- Right to Sell: Consumers are allowed to sell their data to companies but also have the right to revoke such a permission and request the deletion of such data, even if it has been sold to another company.

- Right to Multiple Request Mechanisms: Businesses are required to offer at least two different channels for consumers to request disclosure about the data collected from them, including a toll-free number.

- Right to No Discrimination: Organizations are not allowed to discriminate consumers if these try to exercise their right under the CCPA. However, companies are allowed to change their service level or the price of a good or a service if a data collection is not permitted by the user, provided the price of service level is in relation to the value of the personal data.

- Right to Access: Consumers have the right to request organizations to disclose what data is stored about them, the sources, the use of that information, and what data was sold.

- Right to Opt-out: Businesses are required to set up a Do not Sell my Information button on their website and procedures to comply with the law.

Obviously, there are a few differences. The GDPR requires an Opt-In, whereas this is only required for minors under the CCPA. The GDPR enables consumers to have their data corrected, such a possibility does not exist in the CCPA. Similarly, only under the GDPR, users can object profiling and automated data processing. The CCPA has a strong emphasis on the rights regarding data sale which is not the case for the GDPR.

11.1.2   Do Users Really Care?

Today, data is collected faster, in larger quantities, and possibly stored much longer than users may realize. The consequences of one's own lax handling of data are abstract in most cases. Society, politicians and legislators may be unaware of the long-term consequences of these new data collections and the opportunities and threats they create, especially since the technology for collecting and analyzing data is developing faster than we and, in particular, the legal system can keep up.

Most people don't take privacy information seriously enough as shown by the example of the general terms and conditions of a WIFI operator: By giving their consent, users agreed to clean the toilets at a music festival; 22,000 people either had no problem with this, or did not read through the terms and conditions [13].

What can happen when companies have our data? A well-known phenomenon enabled by profiling is differential pricing, where some users are shown higher prices than others based on their profile [17]. Deleting cookies helps, is the common advice, but nowadays cookies are not even necessary to be able to track a user. Datta et al. published a remarkable article on this in 2015 [19], describing experiments where drug abuse websites were visited, whereupon the ads displayed changed, but not the Google Ad Settings.[1] Worse, when a profile was changed from "male" to "female," fewer ads for high-paying jobs were visible compared with before.

Internet providers store for several months which IP address we use to surf the net. In the event of a legal case, the person behind an IP address may be disclosed to the investigating authorities. But, not everyone is aware of these data collections, because not everyone is aware of the technical possibilities that exist today to identify a supposedly anonymous person after all. Not only do we willingly leave a data trail by being on Facebook, etc., but we also leave

---

[1] A Google site that allowed users to review their ad prefeences.

a data shadow, meaning that we leave behind data that we may not even know is being collected.

Each of us has something to hide, even if you might not believe it at first.[2] Things can go wrong quickly when data falls into the wrong hands. This happened in October 2020, when psychotherapeutic medical records were hacked in Finland, including those of high-ranking politicians [54].

Data does not necessarily have to be stolen by external hackers; the danger can also lurk from within. For example, the robo advisor Scalable Capital reported in October 2020 that customer documents had been accessed with the help of internal company knowledge [39]. And, for those who think these are all exceptions: Google also had a leak in which some of the photos stored in the Google Cloud went into the libraries of other users [11].

11.2 Ethics: Against Profiling and Discrimination

Protecting data is one thing. But even when all the data has been legally collected, the team's work is not yet finished. And by team, I don't just mean the data scientists themselves, but depending on the size of the company, all the stakeholders in a project. As soon as it becomes public that a company's algorithm is used for profiling purposes or discriminates, the company's reputation is at stake and with its survival. Not every company has officers responsible for ethics (sometimes operating under the term compliance), and it is not uncommon for them to lack the qualifications to test the impact of an algorithm. Therefore, everyone on the team is responsible for preventing profiling and discrimination.

11.2.1 What Is Discrimination?

Algorithms can make discriminatory decisions that a human would (hopefully) never make, because these decisions are not correct or even violate laws. But, how can an algorithm discriminate? After all, an algorithm has no biases that a human could have. However, disadvantages can arise from the training data and from the algorithm's weighting. The problem is not the algorithm itself, but the fact that discrimination exists in the world. Data from a world where discrimination exists will accordingly contain that discrimination, so an algorithm will pick it up. Unfortunately, we can find some examples of this in the recent past. In November 2019, *The New York Times* published the case of a software developer, David Hansson, inventor of Ruby on Rails and founder of Basecamp, who got much higher credit limit on his Apple credit

---

[2] If you really think that you have nothing to hide, please send me your passwords so that I can doublecheck :)

So nobody understands THE ALGORITHM. Nobody has the power to examine or check THE ALGORITHM. Yet everyone we've talked to from both Apple and GS are SO SURE that THE ALGORITHM isn't biased and discriminating in any way. That's some grade-A management of cognitive dissonance.

Tweet übersetzen

12:26 vorm. · 9. Nov. 2019 · Twitter for iPad

FIGURE 11.1
Tweet from David Hansson.

card than his wife [51], see also Figure 11.1. However, his wife had a better "credit score". In other cases, too, women who earned more money than their husbands are said to have received a lower credit limit.

The Apple credit card is issued by Goldman Sachs, so Apple is not itself the credit card issuer. The problem, not unique to this case, is that customer service from both Apple and Goldman Sachs insisted that the algorithm does not bias or discriminate, but the algorithm cannot be verified either. The entire thread on Twitter can be read at https://alby.link/applecreditcard.

11.2.2 How to Prevent Discrimination

There is a simple way to test whether an algorithm discriminates or not. The previous credit card example involuntarily contains a possible test, because it would have been sufficient to test whether a woman, just because she is a woman, receives a different decision than a man. In this case, the test goes even further, since the woman had a better credit score than the man. But, it would be enough to compare two completely equal identities that have only one difference: One is a man, the other a woman. Both earn the same money; both have the same address. In this way, it would be possible to find out very quickly whether the algorithm discriminates against the woman only because she is a woman. This approach can also be applied to other characteristics that could potentially lead to discrimination.

However, it is not a good idea to solve the problem in such a way that features such as gender are simply not used, even though in the USA, for

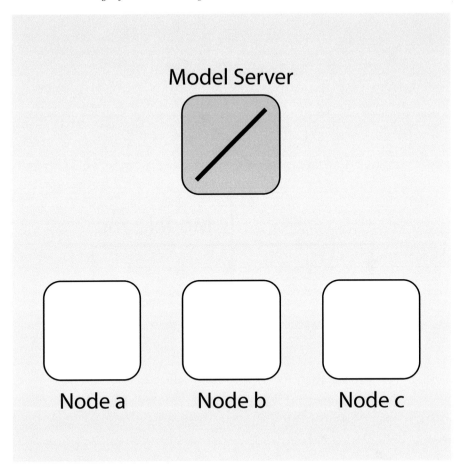

FIGURE 11.2
The server selects a static model for training.

example, gender may not be included in the algorithm [32]. This is because other features may also be influenced by gender, so these features
ensure that gender still matters. A somewhat far-fetched example is shoe
size, which of course, should not play a role in credit applications. But if
the gender variable is not included in the model, shoe size can still ensure that women are treated differently, since most women tend to have
smaller feet than men. If the pay gap is taken into account, i.e. that
women earn less than men in the same position, not even shoe size needs
to be used to account for this. It is necessary to examine the individual
features to see if they function as a proxy for a feature that could be
discriminatory.

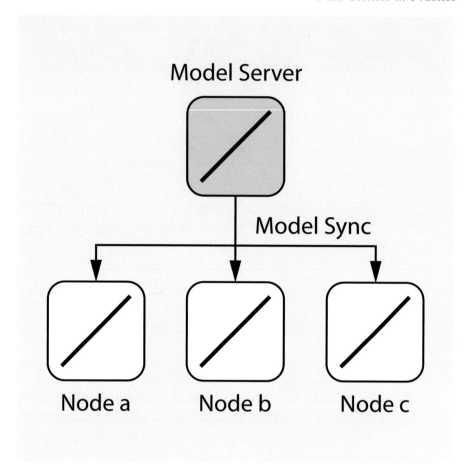

FIGURE 11.3
The model is distributed to the nodes without training.

11.2.3   What Is Profiling?

While probably everyone can imagine something under discrimination, the situation with profiling is different. The GDPR, Article 4, defines profiling as follows:

> [...] any automated processing of personal data which consists in using such personal data to evaluate certain personal aspects relating to a natural person, in particular to analyze or predict aspects relating to that natural person's performance at work, economic situation, health, personal preferences, interests, reliability, behavior, location or change of location

Article 22 then clarifies that profiling is not permitted in the first instance:

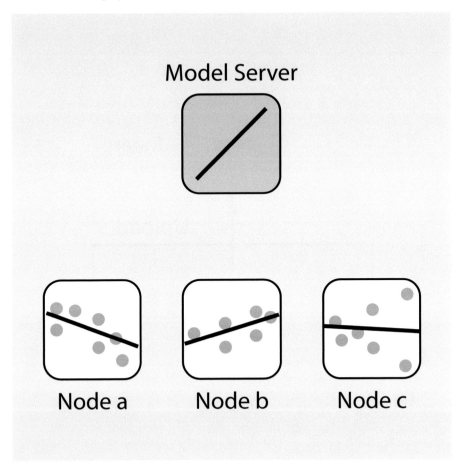

FIGURE 11.4
The nodes each use local data to train the model.

The data subject has the right not to be subject to a decision based solely on automated processing, including profiling.

An example of unauthorized profiling is when the work of employees in a call center is monitored and automated decisions are made based on this. In this example, it is assumed that the employees have given their consent for this data to be collected. For example, the number of calls answered, caller satisfaction, number of sick days, etc., could be collected for each employee and converted into a performance score that would automatically decide on a salary adjustment. This type of profiling is obviously not acceptable. It would be different with a human being between the score and the decision, provided that he can really decide something himself and does not just communicate the

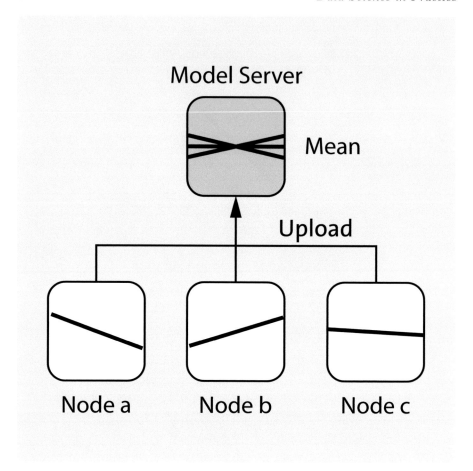

FIGURE 11.5
Global Mode.

decision of the algorithm. Similarly, data from a credit agency may not be used to make automated decisions about the granting of credit or the willingness to enter into a contract. Here, too, there are exceptions, for example, if an automated decision is necessary for the conclusion or fulfillment of a contract, a legal provision exists, or the data subject has given his or her consent.

Although profiling also takes place in the case of personalized online advertising, it is not covered by the ban, because it has no legal effect and no significant consequences arise. However, it can be seen in the example of Datta et al [19]. that negative consequences can also arise here, which may not have been considered in the legislation. Here, too, it can be seen that legislation does not develop as quickly as reality. In the context of advertising, there is another example that has achieved a certain sad notoriety in Data Science circles: It is about the prediction of an algorithm that was used at the supermarket chain Target to identify pregnant women based on their shopping habits [22]. A high school student received off-line advertisements for pregnant women, which prompted her father to complain to Target because he was unaware of his daughter's pregnancy and feared that Target would influence her unfavorably. A short time later, it then became apparent that the young lady was indeed pregnant, but had not yet told her parents. Based on shopping behavior alone, the Target algorithms had determined a high probability of pregnancy, which had probably put the student in a difficult situation with her parents.

Kosinski, Stillwell, and Graep described in an article in 2013 that Facebook-likes reveal much more about a person than initially suspected [33]. They demonstrated that in addition to sexual orientation, religious views, political preferences, and the use of narcotics, it was even possible to determine whether someone had been affected by their parents' divorce at a young age. While the Cambridge Analytica scandal has brought these possibilities further into the public's awareness, nothing seems to have changed in the way Facebook is used. Arguably a bit more tangible than clicking Likes and its consequences is another type of profiling: predictive policing, also known as PredPol. It's about predicting crimes before they happen. Again, historical data is used to make predictions. It is not unlikely, to say the least, that as a result, certain places in a city will be more heavily scrutinized, and also certain types of people, is not unlikely, to say the least. But therein lies the danger of a self-fulfilling prophecy, because with the increased patrols, more crimes could also be detected. Of course, more crimes might actually take place here than elsewhere, but just because you live on a certain street or your parents come from another country doesn't mean you're prone to being a criminal.

Police in Chicago went even further [28] and created a "heat list" on which people were placed who did not even necessarily have to have committed a crime themselves, but were suspected of being capable of committing a crime at some point simply because they had contact with a delinquent. People on the heat list were then warned that they were on that list and that they had better be careful not to do anything else that could get them in trouble. Again, data from social networks was used.

11.2.4 How Can Profiling Be Prevented?

Two approaches can prevent profiling while providing more privacy: differential privacy and federated learning. Differential privacy tries to ensure that an individual's data does not affect the end result. It is based on the fact that noise, meaningless data, so to speak, is injected into a data set that can be removed at an aggregate level. An example by Kelleher and Tierney works by having survey respondents answer a delicate question such as "Have you ever committed a crime?" according to the following principle:

1. Flip a coin and keep the result to yourself.

2. If you have a number, answer "Yes."

3. If you have a head, tell the truth.

The probability for heads or tails is $\frac{1}{2}$ in each case. I assume that you have not yet committed a crime, but you would still answer with a "yes" if you have a number. So 50% of the participants would already answer with a "yes" whether theyy have committed one or not. With the other 50%, however, we learn the truth, namely exactly the proportion of those who have committed a crime. Of course, we don't know who rolled heads or tails, because then we could infer a single person, at least in the case of the heads-throwers. And, we therefore don't know afterwards whether someone has told the truth or not, but nevertheless, we know the proportion of those who have committed a crime.

In federated learning, data from a node, such as a smartphone, is not uploaded to a central server for evaluation; instead, an initial model is trained using local data on the device, and then, the results of the model are transmitted to a central server. The steps in detail:

- Step 1: The server selects a static model for training.

- Step 2: The model is distributed to the nodes without training.

- Step 3: The nodes each use local data to train the model.

- Step 4: The training results are aggregated on the server, which builds a global model from them without having seen the local data.

Apple, for example, is one of the users of this approach (see especially [10]). Allegedly, even Siri is trained in such a way that its own speech data is no longer uploaded to a central server, but the speech recognition is optimized locally. Google also uses this approach for the Google Keyboard on Android.

12

Next Steps after This Book

You'll never walk alone. (Gerry and the Pacemakers)

You have worked through the entire book – congratulations! Now, you're probably wondering how you can further quench your thirst for knowledge. That's what this chapter is all about.

12.1 Projects, Projects, Projects

Practice makes perfect; but where do you get projects if you don't have a job as a Data Scientist? In fact, many opportunities exist to dive deeper into the subject matter. Regular practice is necessary because every project, every question, and every data set is different. Accordingly, you can learn on every project. No book, course of study, or tutorial in the world will train you to be a good Data Scientist; it's all about practice. But – where to start?

12.1.1 Putting Together a Project Portfolio

First of all, many problems in this book allow further steps. For example, what happens when you use different parameters in a training? Why does it happen? Understanding this will take you much further than any example that you simply reprogram.

Also, the many data sets that R comes with out of the box invite further sessions with R and independent experimentation. With the following command, you will get an overview of all data sets that are already available in R:

```
library(help = "datasets")
```

Not all data sets are useful for machine learning, but each is intended for specific tasks. Taking arules as an example, some packages also bring their own data sets.

Another way to get more involved with Data Science is to keep your eyes open to what problems or issues exist around you. There are plenty of problems; the only question is which one is worth solving. The following examples are from my project portfolio:

DOI: 10.1201/9781003426363-12

- When the Corona crisis ran its course, the media was full of data, mostly from Johns Hopkins University. Not all of the data visualizations were useful to me, so the decision was easy to build a dashboard myself. This is not so much an example of Data Science as it is a good exercise in how to merge, transform, and deliver disparate data in an interactive Shiny app. I shared the dashboard on social media and then received good feedback on what features were missing, which I then developed in further steps.

- A few years ago, I wanted to buy a used SLR camera, namely the Canon 5D Mark III, which was still popular at the time. But, the prices quoted in the classifieds were so different that I was unsure what was really a reasonable price. With SLR cameras, the shutter release wears out with the number of releases and must be replaced at some point. Depending on the model and usage, this may be sooner or later. With professional cameras, which are used by professional sports photographers, for example, you quickly reach several hundreds of thousands of releases with used cameras. Here, the probability that the shutter release will not last long is relatively high, even if the camera is not very old. So, the number of releases should have an influence on the price, besides the accessories, and very likely, this is a linear function that can be identified with a linear regression. You can see the result of the analysis at that time in Figure 12.1. While the data collection is quite small, and heteroskedasticity is obvious, nevertheless, the model helped immensely in my purchase decision. For example, since the used cameras with a high deviation from the regression line did not include any accessories that justified the deviation, it was clear to me that the sellers had unrealistic expectations. I therefore decided to buy a used camera whose price was close to the regression line. Accessories, by the way, play less of a role in pricing than previously thought. You can read the whole article on my blog at https://alby.link/regression.

- I had always wondered what criteria are used at online news sites to decide whether an article is behind a paywall or not. In this case, I (very carefully, of course) "scraped" some article headlines and teasers (with Python, because it was really easier) and marked what was a pay article and what was not.

- In the next step, I developed a classifier that can predict whether an article will become a pay article or not based on the words in the headline and teaser. In fact, this has worked very well, with new articles having to be added to the corpus every so often to keep the model up to date. Corona was not a topic in 2016, so my model could not handle it initially. In addition, I created word clouds to see what the differences in topics were in the two groups. Besides brushing up on various natural language processing (NLP) techniques, what did I take away from this project? I radically reduced my time on news sites.

  You may think of these little projects as gimmicks or tinkering, but they are

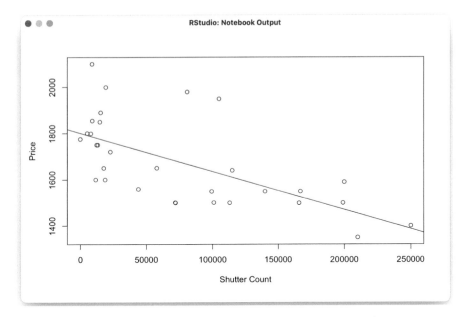

FIGURE 12.1
Linear regression to identify the appropriate price for a used Canon 5D Mark III.

wonderful finger exercises, also for trying out new libraries. Most importantly, they lead from a theoretical knowledge to a practice. A potential employer is more interested in what you've actually already done, not what you've read somewhere and what course you've taken.

A private project portfolio is also useful when little or nothing can be said about work projects. For example, I've developed some exciting things in my career, but I'm not allowed to talk about them because of confidentiality agreements. Some of these projects, however, had side aspects that were not covered in the project, but which I nevertheless found so interesting that they then became private projects.

What is very important in these projects is that no false respect for the task keeps you from just giving it a try. Often, the things that seem difficult are much easier than you think, and the things that seem easy cost much more time than planned. Again, experience helps to make more accurate estimates in the future.

12.1.2 Kaggle

Kaggle was founded in 2010 and has been running competitions for data scientists ever since. Kaggle has been part of the Google family since 2017. Some of the contests come with substantial prize money. For example, Kaggle and

Google offered a competition to classify YouTube videos, which came with 100,000 dollars in prize money, the first prize was 30,000 dollars. Netflix even offered 1,000,000 dollars in a contest.

But, unlike the famous Frank Zappa song, data scientists aren't just in it for the money. Some contests focus on how to better detect certain diseases. Among other things, a contest existed using data sets with thousands of photos of moles to train whether skin cancer could be better detected with algorithms.

Kaggle also offers a wonderful number of data sets to work on outside of the competitions. In addition, other data scientists share their notebooks so that you can also learn from them. However, caution is advised here as well, because not every user of Kaggle is an experienced data scientist. A look at the profile shows how experienced the contributors really are. There are additional discussions on the notebooks where models and approaches are discussed.

The default language on Kaggle is Python, but notebooks can be switched to R (see Figure 12.2). Jupyter notebooks are used, which I mentioned in Chapter 5, "Exploratory Data Analysis". Combining both languages in one notebook is not possible in a Jupyter notebook. Jupyter is not as comfortable in all aspects as the notebooks in RStudio, also the keyboard combinations are different, so some getting used to it is necessary. But, the computing power provided is considerable; at the time of completing this manuscript, for

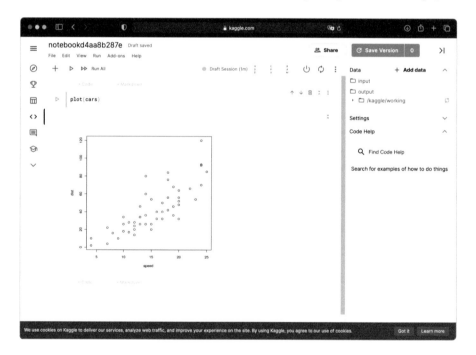

FIGURE 12.2
Kaggle also offers R.

example, 16 GB of RAM is available, twice as much as in Posit Cloud, and that for free. It is also possible to add a GPU[1] as an "accelerator".

12.2 Where to Find Help

12.2.1 RTFM

A typical response of a nerd to a question from a user is "RTFM", an acronym that stands for "Read The Fantastic Manual" (well, not quite, but surely you guess what the F stands for anyway). The manual is first of all, the help for a command, which is started with ?command, as already described in chapter 4, "Introduction to R".

For many packages there are additional vignettes that provide detailed documentation. The command

```
browseVignettes()
```

lets you display the installed vignettes. If you specify the name of the package plus quotes in the parenthesis, exactly this vignette will be displayed, for example:

```
browseVignettes("arules")
```

Not all vignettes are of high quality: in some cases, there is nothing more in them than in the help, while others are almost scientific papers that describe the approach of the package in great detail. The vignette on arules, for example, is a positive highlight, currently a 38-page document that introduces the topic of association algorithms in its entirety. It is therefore much more than just a guide to the package itself.

Another possibility is to search for individual terms or concepts, for example with

```
help.search("tibble")
```

With

```
RSiteSearch("tibble")
```

the whole r-project.org site is searched. The "Introduction to R" [50] is also highly recommended. It is updated continuously.

The biggest challenge in the help and also in the vignettes can be that so many technical terms are used there that you understand even less after

---

[1] Graphics processing units are often for machine learning tasks due to their parallel processing power and high memory bandwidth

reading than before. To cut a long story short: It's not up to you. Not every contributor has mastered communication appropriate to the target audience or believes that getting started should be made as easy as possible for everyone. I should have explained most of the terms in this book.

At a time when the desire for "instant gratification" is increasing and attention spans are decreasing, the recommendation to read the manual first may seem strangely old-fashioned. But, Data Science is not an easy craft to learn; it requires concentration and a lot of time. However, if you have managed to work your way through this book, the help pages and vignettes will be no obstacle to you. Self-study of the help pages is also an absolute prerequisite for the steps presented in the next sections.

12.2.2   Stack Overflow

Many a programmer claims that he owes his career to Stack Overflow, and one wonders how it used to be possible to program without Stack Overflow.

It is unlikely that you have an issue that has never existed before and has not appeared somewhere on Stack Overflow. Unlikely doesn't mean impossible. If you are using new packages or already have a lot of experience, the probability increases, of course. However, two hurdles exist:

- You must ask your question in English.

- You need to do your "homework" in advance.

The community on Stack Overflow is not without its own demands, because people here give you their time even though they barely get anything out of it themselves (more later on what exactly they get). That means if you don't do your homework, you're wasting other users' time, and you'll be resented for it. This means that before you ask a question, RTFM!

As a second step after reading the help or manual, it is absolutely necessary that you first thoroughly research whether a similar or possibly even the exact same question has already been asked on Stack Overflow. Otherwise, your question will be closed and marked as a duplicate. This is not a nice feeling. It feels even worse when your question is marked as useless and a minus appears next to the score. In the third step, you should make sure that you are using the right division of Stack Overflow. There exists not only the main page stackoverflow.com, but also sub-pages for almost every topic, for example:

- Cross Validated for Statistics, Data Mining, Data Analysis and Machine Learning at https://stats.stackexchange.com

- Data Science at https://datascience.stackexchange.com for Machine Learning

- Artificial Intelligence at https://ai.stackexchange.com for everything about artificial intelligence

FIGURE 12.3
The anatomy of a stack overflow question page.

- Computer Science at https://cs.stackexchange.com for those interested in computer science

The individual sections do not overlap, as you can already see here. The search on stackoverflow.com searches all sites, so this should be the start. You can see the complete overview of all sites at https://stackexchange.com/sites. The search should ideally also include the programming language for which you are looking for a solution, because data frames and other concepts also exist in Python and other languages.

The fourth step is that you provide a reproducible example. Describe your question as precisely as possible, and provide your previously used code and data. Of course, you must not reveal any sensitive data; dummy data is enough, and only a few lines of it, so that the Stack Overflow community can and will help you quickly. The data and your code should be easy to copy out so that anyone can immediately reproduce the example in their own R installation. Ideally, you should also use your example data to show what the desired result should look like. Language is ambiguous; an example can speed up understanding.

Your example, by the way, is entered in a kind of Markdown language that is very close to what you already learned about in Chapter 5, "Exploratory Data Analysis."

Figure 12.3 is an example of a Stack Overflow page that contains a few interesting details. First, to the left of the question text is a number; it indicates how often people voted for or against this question. Clicking on the arrow above the number expresses that this question has been preceded by research effort and is clear and useful. Clicking on the arrow below the number means the opposite. So, the questioner is given immediate feedback on how good the question is and whether the homework has been done. If your question immediately lands in the negative range, then you know that you need to invest more time in your homework next time. You may also find that your question is closed, either because the same question has already been asked elsewhere or because it elicits responses that may be based more on opinion than fact.

For logged-in users, further data can be seen at the top right. First of all, a number that expresses the reputation of a user. There are points for various actions, for example:

- for a question that is looked at by many users

- for a question that is "upvoted" by many users

- for an answer that was marked as helpful

Depending on how many points a user has, he or she gets privileges, for example, to see less advertising or to vote questions up or down. Conversely, this means that you have to have published a few good questions or answers before you can actively participate in Stack Overflow. Likewise, you can lose reputation; there are also negative points for bad answers. So, the Stack Overflow system is very smart, because it gives more authority to those who ask meaningful questions and give good answers. It is difficult if not impossible, to "fake" your way to the top. It's normal to feel ashamed of the questions you asked in the beginning after a few years. That's okay. Everyone starts out once. But, please don't make beginner mistakes like not doing enough research in advance or not reading the manual.

12.2.3   The R-Help Mailing List

The mailing list may seem antiquated next to Stack Overflow, but it is not. Here, too, a few questions come together every day, and R experts respond (see figure 12.4).

Another big advantage of the mailing list is that you're not exposed to ads, which you'll definitely see on Stack Overflow. And where Stack Overflow is more of a pull medium – that is, you use it when you need it – the mailing list brings you new questions every day that you had never asked yourself before, and you learn along the way. The effect is not to be underestimated, because you expand your horizons and learn new methods.

The R-Help mailing list also has its rules, and most of Stack Overflow's can be adopted here. However, you will not be resented if a question has been asked

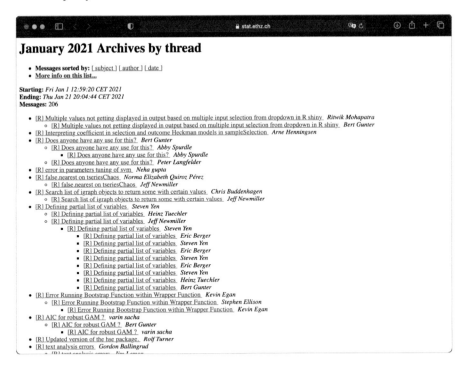

FIGURE 12.4
Just like old times: the R-Help mailing list.

before. But, the scope of questions is limited: statistics questions will only be answered if you formulate them really well and they are also interesting for the rest of the group. There are separate mailing lists for statistics questions in general. The mailing list is also not suitable for getting tasks for school or university done by others. Likewise, questions about platforms like macOS or packages should be asked in the respective help forums. The rules for good questions look something like this:

- The subject should be specific enough that the list participants know what it is about.

- All "homework" should have been done in advance, i.e. everything I have already mentioned.

- There should be more in the mail than "this command does not work".

- There should also not be too much in the mail, but most importantly, the goal you are actually pursuing should be clear.

- In certain cases, when unexpected behavior is involved, you should also mention the version number of R and the operating system.

- Please install the latest version of R, or check at least whether your issue is already known or fixed in a current version.

You can find the mailing list sign-up page at https://stat.ethz.ch/mailman/listinfo/r-help.

12.3   RSeek

RSeek is a search engine built specifically for R. It can be reached at https://rseek.org. Unfortunately, it is not transparent how often it crawls R-related sites and how the ranking works. But from my own observation, I can say that the most important R sites are covered. Apparently, the search is based on a Google Site Search.

13

Appendix: Troubleshooting

13.1 Typical Error Messages and Solutions

R is unfortunately not famous for displaying particularly helpful error messages. Therefore, in this section, we will cover some common errors and discuss them in detail.

13.2 Typical Mistakes and How to Avoid Them

The most common type of error is: typos! But, other errors are also covered in this list:

- Variable or function name misspelled: This is actually the most common error I see in the seminar room.

- Case-sensitive: R is case-sensitive, so library() works, but Library() spits out an error message.

- While R can handle both American and British English, as we see with summarize() and summarise(), no function likes to be misspelled. Unfortunately, there is no spell checker, but it should give you pause if R makes no suggestion after you type three characters.

- Did they forget a closing parenthesis or closing quotation mark?

- Missing comma or missing pipe: very common error, the first one mainly in Shiny applications, the second one in dplyr code. If the pipe is missing (%>%), the rest of the code will not be executed. Bonus error: If there is also a pipe after the last line, then an error message may also occur. Or, nothing happens at all.

- Library not loaded. Also a very common error. Better load once more than once less.

- Object not defined or loaded: If an object is in memory because it was defined on the console but not in the notebook, it will not be present the next time R or RStudio is loaded. This causes the first error. But, it can also happen that something has already gone wrong when defining the object. To fix such a problem, it makes sense to test one line of code at a time, rather than running an entire script. If a program in the dplyr dialect consists of 10 lines, run the first line first, then the first two, then the first three, and so on until you have identified the line where the problem occurred. This is where another advantage of the Tidyverse reveals itself: With complex expressions in Base R, it can be immensely hard to find the error. By dividing them into lines, on the other hand, errors can be debugged much more easily.

13.3   R or RStudio Does Not Respond

If R or RStudio does not respond, there can be several reasons for this. The most common reason is that a script is running that blocks R. This can be due to a programming error, but possibly also due to the amount of data you want to process. Another problem can occur when a notebook becomes very large. This error occurred mainly with earlier RStudio versions, but may also be due to the notebook itself.

If all else fails: Every operating system offers the possibility to stop a program. The big disadvantage: Everything you have entered up to that point and not saved is lost. Therefore, you should get into the habit of always having your left hand on Command+S (or whatever your keyboard shortcut is on your operating system) to save, and do so after every working step.

13.4   Typical Error Messages

These error messages are the ones I see most often from participants in my courses:

- could not find function XXX: Most likely, a library has not been loaded. If you have already written your own function, it could also be that it has not been loaded yet.

- could not find function "%>%": As described above. You have not yet loaded the Tidyverse that you apparently wanted to use.

- no applicable method: This error occurs when you want to apply an object-oriented function on an object that is not suitable for it.

- non-numeric argument for binary operator: You have tried to cast an object that does not consist of numbers to a function that takes only numbers.

- object **XXX** not found: You have not loaded the object yet, or it has not been defined at all.

- Cannot open connection: The most common reason for this error is that a file could not be opened, for example, because its path was not specified correctly. This error occurs quite often because the working areas of R and the current working directory are different. Such errors can also occur when a script is transferred from one computer to another.

14

Glossary

A/B test: See Split Testing

API: API stands for application programming interface. Instead of interacting with an application in a graphical user interface (GUI), an application communicates with another application via an API. For example, report data can be retrieved, updated daily, and then mapped to another application.

AUC: Refers to the receiver operating characteristics curve (ROC curve). The larger the area under this curve, the better a model works. See Chapter 3, "The Anatomy of a Data Science Project."

Average/Arithmetic mean/Mean: Mean value where the sum of the values is divided by their number. Susceptible to outliers. See also median and mode.

Average of the sample/Mean (of a sample): The average of a sample differs in that it is not known what the actual average of the population is. This inaccuracy is taken care of by dividing by n-1 and not by n, i.e. the number of elements in a sample.

Bernoulli distribution: A distribution of events where there are only two possible outcomes. The most common example is a coin toss where either heads or tails can come out. This is also known as a Bernoulli trial.

Big Data: Buzzword; Big Data is rarely found in the wild. See Chapter 2, "Machine Learning, Data Science and Artificial Intelligence."

Binary Data: Variables that can take only two values; see also Bernoulli distribution.

Binomial distribution: Special case of the Bernoulli distribution where n Bernoulli trials are performed.

Blind study: If the participants in a study do not know whether they have been given a drug or a placebo, this is called a blind study. If the people treating the participants also do not know which drugs they are administering, this is called a double-blind study.

Bootstrapping: Sampling with back loading.

Categorical Variable: A categorical variable exists when there are two or more categories (for example, diverse, male, and female) but there is no order. Sometimes, such a variable is also called nominal.

Central limit theorem: If samples are taken from any distribution and the arithmetic mean is taken from each of these samples, these means will give a normal distribution, regardless of the distribution from which they originate. Probably one of the most important approaches in statistics, since many other approaches are derived from it.

Confidence interval: Indicates with a probability in which range a parameter of a random variable lies, for example, the arithmetic mean. Sometimes abbreviated to CI, but should not be confused with artificial intelligence.

Confirmation bias: The tendency to seek and perceive only information that corresponds to one's own opinion.

Continuous variable: A variable that can take on any value within an interval.

Control Group: In a split test, half of the test subjects are given a treatment treatment, and the other half are given a placebo.. This second group is also called the control group. In the field of Data Science, a split test is primarily used to see whether the results of the model actually produce significantly better results than no model.

Convenience sampling: Refers to a type of sampling in which, for example, you simply go outside the door and grab a few passers-by to ask them which party they would vote for next Sunday if there were federal elections. Obviously, where you ask has an impact. To get a representative picture, you can't organize a poll out of convenience at the door.

Cross validation: In cross validation, one part of the data is taken for training and another for testing; often, one fifth is used for testing. This process is performed several times, always using different fifths of the whole data set for testing.

Data lake: Data storage concept in which the raw data flows in and is not pre-structured as for a data warehouse.

F distribution: The F distribution is a probability distribution used to compare the variance of two samples. It is only used for testing and is named after Ronald Fischer.

Feature engineering: The application of domain knowledge to the extraction of features suitable for model building.

Heteroskedasticity: See Chapter 6, "Forecasting use case."

Hypothesis testing: See Statistical test.

Mode: The mode is the most frequent value in a data set. There can be more than one mode.

Nominal variable: See Categorical variable.

Normal distribution: The normal distribution is the prerequisite for some statistical procedures, for example, the t-test. Also known as the Gaussian distribution . However, other distributions are not abnormal; the normal distribution is just a name.

Null Hypothesis: In a hypothesis test, there is on the one hand the alternative hypothesis, i.e. what we actually want to prove, and on the other hand the null hypothesis, which claims the opposite, so to speak. We want to disprove the null hypothesis in order to be able to say that our alternative hypothesis seems to be true.

Ordinal Variable: A variable classified on an ordinal scale. Unlike a categorical variable, where the categories have no order, this is different for an ordinal scaled variable. For example, if the categories are school-leaving qualifications, a ranking can be created in this case.

Overfitting: Creating a model that has been adjusted so much to fit the training data that it only performs well on this data.

Placebo: In medicine, a placebo is a drug that has no effect and is used primarily in split tests to see if a drug actually does anything.

Probability distribution: Specifies the probability of each possible value in a random experiment. Example: What is the probability that I roll a 6 one time, three times, or six times in a row? This example can be described with a binomial distribution.

p-value: The probability of seeing the present values if the null hypothesis is true. Suppose the average man is 6 feet tall, and we want to know if people in Berlin are taller. Our null hypothesis is that there is no difference. We go to a mall and ask a group of men if we can briefly check their height. Interestingly, these men are all over 2 meters tall. It is very unlikely that we accidentally asked a basketball team to participate, but it is possible (as in this example). This is exactly the probability that the p-value reflects. In this case, we are saying that it is very unlikely that we accidentally included only very large men in the sample, and we are incorrectly rejecting the null hypothesis. This is also an example of a type 1 error.

Receiver operating characteristic (ROC) curve: Curve that maps the frequency distribution of sensitivity and specificity. Used together with the area under the curve.

Regression to the Mean: The effect that after an extreme value comes again a value that is closer to the average.

Residual: Sometimes called error; difference between an observation and a statistical model.

Sensitivity: Indicates the probability that a test actually detects a disease. See also specificity.

Specificity: Indicates the probability that a test recognizes a healthy person as such. See also sensitivity.

Split testing: Also A/B testing. A group of participants is divided into two subgroups; one is the control group, the other the test group. The test group receives a treatment; the control group receives a placebo. If the treatment is indeed effective, a statistically significant effect should be measurable.

Standard Deviation: In simple terms, the mean deviation of the data points from the mean. In a normal distribution, 95% of the population is within plus/minus 1.96 standard deviations.

Standard error: The standard error indicates how far the sample mean deviates from the actual mean.

Standardization: See z-score.

Statistical Significance: A result is said to be statistically significant if the p-value is below a pre-specified threshold.

Statistical Test: Also called hypothesis test or significance test. also called significance test. Statistical procedure used to draw conclusions about the population on the basis of a sample. As a rule, a hypothesis is created here and its validity or invalidity is determined. A well-known example is the t-test.

Supervised Learning: Less frequently supervised learning. Machine learning approach where already known classes (spam/no spam) are used for training so that they can be automatically detected on new data. Typical use in classification and regression.

t-distribution: Also Student's t-distribution. In the case of a sample, the standard deviation of the population is usually unknown, so it must be calculated from the sample. In this case, the t distribution is used. This is dependent on the number of degrees of freedom (often abbreviated as df in R). In the past, the t-value had to be looked up in tables, but today, the statistics programs display this value automatically, just like the corresponding p-value.

Type I error: An Type I error occurs when a null hypothesis is rejected even though it is actually true. Example: I give out a stat candy to half of the participants in my class. Alternative hypothesis: The participants have greater success on the final exam. Null hypothesis: The candy has no effect at all. Now, in fact, most of the participants who received the candy write a better test, not because the candy had worked, but simply because they had paid better attention. Then, I would falsely reject the null hypothesis and see something that is not there.

Type II error: An type II error occurs when the null hypothesis is falsely retained although the alternative hypothesis is true. To stay with the example about the error of the 1st kind, it could be that my stat candy actually works. The alternative hypothesis is that the stat candy has a positive effect on the test result, and the null hypothesis is that the candy has no effect at all.

Unsupervised learning: Machine learning approach where the different classes are not known, and the algorithm finds patterns independently. An example is clustering.

Z score: The Z score is a standardized value of a variable. For example, if we have a data set of people whose height and weight have been measured, the numerical ranges of these data are very different and not comparable, for example, 1.71 meters versus 68 kilos. If we now subtract the arithmetic mean of each variable from each data point and divide the result by the standard deviation, we get a comparable value. Some machine learning algorithms require standardized values, especially when these algorithms use distances.

Bibliography

[1] Charu C. Aggarwal, Alexander Hinneburg, and Daniel A. Keim. On the Surprising Behavior of Distance Metrics in High Dimensional Space. In Gerhard Goos, Juris Hartmanis, Jan Van Leeuwen, Jan Van Den Bussche, and Victor Vianu, editors, *Database Theory — ICDT 2001*, volume 1973, pages 420–434. Springer Berlin Heidelberg, Berlin, Heidelberg, 2001. Series Title: Lecture Notes in Computer Science.

[2] Rakesh Agrawal, Tomasz Imieliński, and Arun Swami. Mining association rules between sets of items in large databases. *ACM SIGMOD Record*, 22(2):207–216, June 1993.

[3] Tom Alby. *Web 2.0. Konzepte, Anwendungen, Technologien*. Carl Hanser, München, 3 edition, 2008.

[4] Tom Alby. Project Management in Data Science, August 2017.

[5] Tom Alby. *Einführung in die Webanalyse*. Rheinwerk Computing, 2019.

[6] Tom Alby. Data Science: Von der Sprache der Daten zur Sprache der Algorithmen. *API Magazin*, 2(1), 2021.

[7] Andreas C. Müller and Sarah Guido. *Introduction to Machine Learning with Python. A Guide for Data Scientists*. O'Reilly, 2016.

[8] F.J. Anscombe. Graphs in Statistical Analysis. *American Statistician*, 27(1):17–21, 1973.

[9] Vic Barnett. The Study of Outliers: Purpose and Model. *Journal of the Royal Statistical Society. Series C (Applied Statistics)*, 27(3):242–250, 1978. Publisher: [Wiley, Royal Statistical Society].

[10] Abhishek Bhowmick, John Duchi, Julien Freudiger, Gaurav Kapoor, and Ryan Rogers. Protection Against Reconstruction and Its Applications in Private Federated Learning. 2019.

[11] Thomas Brewster. Google Guilty Of 'Big Screw Up' That May Have Leaked Your Videos To A Random Stranger, April 2020.

[12] Jenny Bryan and Jim Hester. Happy Git and GitHub for the useR.

[13] Sophie Burfeind. 22 000 Menschen willigen ein, Klos zu putzen, July 2017.

[14] Casey Ross and Ike Swetlitz. IBM's Watson supercomputer recommended 'unsafe and incorrect' cancer treatments, internal documents show, July 2018.

[15] Chris Chapman and McDonnell Feit Feit. *R for Marketing Research and Analytics.* Use R! Springer International Publishing, 2015.

[16] Tianqi Chen and Carlos Guestrin. XGBoost: A Scalable Tree Boosting System. *CoRR*, abs/1603.02754, 2016.

[17] Stephanie Clifford. Shopper Alert: Price May Drop for You Alone. *The New York Times*, August 2012.

[18] Cole Nussbaumer Knaflic. *Storytelling with Data: A Data Visualization Guide for Business Professionals.* Wiley, 2015.

[19] Amit Datta, Michael Carl Tschantz, and Anupam Datta. Automated Experiments on Ad Privacy Settings: A Tale of Opacity, Choice, and Discrimination. *Proceedings on Privacy Enhancing Technologies*, 2015(1):92–112, 2015.

[20] Thomas H Davenport and D.J. Patil. Data Scientist: The Sexiest Job of the 21st Century., 2012.

[21] Dheeru Dua and Casey Graff. UCI Machine Learning Repository, 2017.

[22] Charles Duhigg. How Companies Learn Your Secrets. *The New York Times*, February 2012.

[23] Ludwig Fahrmeir, Rita Künstler, Iris Pigeot, and Gerhard Tutz. *Statistik. Der Weg zur Datenanalyse.* Springer, 2011.

[24] Andy Field. *Discovering Statistics using R.* Sage, 2012.

[25] Michael Fitzgerald. *Einstieg in Reguläre Ausdrücke.* O'Reilly, 2012.

[26] John W. Foreman. *Data Smart: Using Data Science to Transform Information into Insight.* Wiley, 2013.

[27] Gartner. Understanding Gartner's Hype Cycles, August 2018.

[28] Jeremy Gorner. Chicago police use 'heat list' as strategy to prevent violence. *Chicao Tribune*, August 2013.

[29] Kevin Gray. Kobalt changed the rules of the music industry using data – and saved it. *Wired*, May 2015.

[30] Garret Grolemund and Hadley Wickham. *R for Data Science.* O'Reilly, 2016.

[31] John D. Kelleher and Brendan Tierney. *Data Science.* The MIT Press, 2018.

[32] Will Knight. The Apple Card Didn't 'See' Gender—and That's the Problem. *Wired*, November 2019.

[33] Michal Kosinski, David Stillwell, and Thore Graepel. Private traits and attributes are predictable from digital records of human behavior. *Proceedings of the National Academy of Sciences of the United States of America*, 110(15):5802–5805, April 2013.

[34] Ray Kurzweil. *How to create a mind. The secret of human thought revealed.* Penguin Books, 2012.

[35] David Loshin. Data Quality and MDM. In *Master Data Management*. Morgan Kaufmann, 2019.

[36] Miriam Fauzia. Fact check: Facebook didn't pull the plug on two chatbots because they created a language, July 2021.

[37] Zainab Mudallal. Airbnb will soon be booking more rooms than the world's largest hotel chains, January 2015.

[38] Nitasha Tiku. The Google engineer who thinks the company's AI has come to life. *The Washington Post*, June 2022.

[39] Olivia von Westernhagen. Scalable Capital: Robo-Advisor meldet unbefugten Zugriff auf Kundendaten. *heise online*, October 2020.

[40] Stefan Papp, Wolfgang Weidinger, Mario Meir-Huber, Bernhard Ortner, Georg Langs, and Rania Wazier. *Handbuch Data Science. Mit Datenanalyse und Machine Learning Wert aus Daten generieren.* Hanser, 2019.

[41] Project Management Institute, editor. *A Guide to the Project Managemenet Body of Knowledge (PMBOK Guide)*. The Stationery Office Ltd, Pennsylvania, 6 edition, 2017.

[42] Foster Provost and Tom Fawcett. *Data Science for Business: What you need to know about data mining and data-analytic thinking.* O'Reilly, 2013.

[43] Daniel Rosenberg. Data before the fact. In Lisa Gitelman, editor, *Raw Data is an Oxymoron*, Infrastructures. The MIT Press, 2013.

[44] Sebastian Sauer. *Moderne Datenanalyse mit R: Daten einlesen, aufbereiten, visualisieren, modellieren und kommunizieren.* FOM-Edition. Springer Fachmedien Wiesbaden, Wiesbaden, 2019.

[45] Colin Shearer. The CRISP-DM model: the new blueprint for data mining. *Journal of data warehousing*, 5(4):13–22, 2000.

[46] Andrew S. Tanenbaum. *Moderne Betriebssysteme.* Pearson Studium, München, 2009.

[47] John W. Tukey. The Future of Data Analysis. *The Annals of Mathematical Statistics*, 33(1):1–67, March 1962.

[48] John W. Tukey. *Exploratory Data Analysis.* Addison-Wesley Series in Behavioral Science. Pearson, 1977.

[49] A. M. Turing. Computing Machinery and Intelligence. *Mind*, LIX(236):433–460, October 1950.

[50] W.N. Venables, D.M. Smith, and and the Core Team. An Introduction to R. Notes on R: A Programming Environment for Data Analysis and Graphics., October 2020. Version 4.0.3.

[51] Neil Vigdor. Apple Card Investigated After Gender Discrimination Complaints. *New York Times*, October 2019.

[52] Edgar P. Vorndran. *Entwicklungsgeschichte des Computers.* VDE-Verlag, Berlin, Offenbach, 2. erweiterte auflage edition, 1986.

[53] Thomas Jr. Watson. *Der Vater, der Sohn und die Firma.* Heyne, 1999.

[54] William Ralston. They Told Their Therapists Everything. Hackers Leaked It All. *Wired*, 2021.

[55] Eliezer Yudkowsky. Artificial Intelligence as a positive and negative factor in global risk. In *Global Catastrophic Risks.* Oxford University Press, July 2008.

[56] Aeberhard, Stefan and Forina, M. (1991). Wine. UCI Machine Learning Repository. https://doi.org/10.24432/C5PC7J.

Index

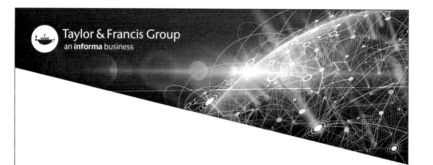